Lecture Notes on Coastal and Estuarine Studies

26

Eckart H. Schumann (Ed.)

Coastal Ocean Studies off Natal, South Africa

Springer-Verlag
New York Berlin Heidelberg London Paris Tokyo

Editor

Eckart H. Schumann
Department of Oceanography, University of Port Elizabeth
P.O. Box 1600 Port Elizabeth 6000, Republic of South Africa

ISBN 3-540-96895-4 Springer-Verlag Berlin Heidelberg New York
ISBN 0-387-96895-4 Springer-Verlag New York Berlin Heidelberg

Printed in Germany

Printing and binding: Druckhaus Beltz, Hemsbach/Bergstr.
2837/3140-543210 – Printed on acid-free paper

PREFACE

Oceanographic research has expanded rapidly in the latter half of this century. To a large extent this has been as a result of an increased awareness of the importance of the ocean, while new technologies and resources have enabled a much better data coverage to be obtained.

Lying along the eastern seaboard of South Africa, the province of Natal has many important associations with the ocean. Oceanographic research here has gone through several phases, without apparently a definite commitment on the part of the authorities to support such investigations. At present there is limited work being done, without much prospect for a major upturn in the foreseeable future.

Nonetheless, there is a considerable wealth of information available on the region and in this volume specialists from academia and state-funded research organizations present analyses of such studies made off Natal. This specifically covers the coastal ocean area, defined as lying beyond the surf zone, but not extending into the deep ocean. Obviously, though, it is not possible to demarcate a natural environment so specifically, and influences from beyond these artificial limits also play a significant part in building up an understanding of the whole system.

As is characteristic of such a multi-authored volume, considerable effort was expended on reviews of the individual chapters. I am grateful to the following individuals for the time and effort they put into this process, and know that the authors of the various chapters benefited from this additional input of knowledge and advice:
G B Brundrit, P Chapman, A J de Freitas, R V Dingle, M L Grundlingh, P A Hulley, L Hutchings, J R E Lutjeharms, A F Pearce, K S Russell, I C Rust, R W Shone, D H Swart, J H Wallace, T H Wooldridge. I am also grateful for the advice and encouragement given to me in the initial stages by A E F Heydorn and F P Anderson, Director and Chief Director, respectively, of the National Research Institute for Oceanology.

Many of the investigations reported in this volume were carried out using the RV Meiring Naudé, the research vessel owned and run by the

Council for Scientific and Industrial Research (see the Appendix). Close co-operation occurs between the ship's crew and the scientific complement during such cruises, and it is a pleasure to acknowledge the agreeable atmosphere that exists on board the RV Meiring Naudé; we are indebted to Captain G A E Foulis and his crew for help received during many hours at sea. Sophisticated data collection techniques also require considerable development and maintenance, and the expertise in the Electronics and Instrumentation Group, under the direction of C C Stavropoulos, contributed greatly to the successful collection of data.

It is often the case that scientists need advice in their communication via the written word; Felix Lancaster reviewed the manuscripts, and his suggestions have helped to make the presentations more lucid, and hopefully more readable. Finally, production of a camera-ready manuscript involves a great deal of manipulation of figures and tables, typing and re-typing, to see that it all fits together. Jean Harris was responsible for this task, and I am indebted to her for making sure that this volume was finally completed.

Eckart H Schumann

CONTRIBUTORS

Robin A Carter, National Research Institute for Oceanology,
CSIR, P O Box 320, STELLENBOSCH, 7600 CAPE.

Allan D Connell, National Institute for Water Research
CSIR, P O Box 17001, CONGELLA, 4013 NATAL.

Jeannette d'Aubrey, National Research Institute for Oceanology
CSIR, P O Box 17001, CONGELLA, 4013 NATAL.

Burghard W Flemming, Forschungsinstitut Senckenberg,
Schlensenstrasse 39a, 2940 WILHELMSHAVEN, WEST GERMANY.

E Rowena Hay, Department of Geotechnology,
Private Bag X256, PRETORIA, 0001 TRANSVAAL.

Ian T Hunter, National Research Institute for Oceanology,
CSIR, P O Box 320, STELLENBOSCH, 7600 CAPE.

Desmond A Lord, Department of Oceanography, Univ. of Port Elizabeth,
P O Box 1600, PORT ELIZABETH, 6000 CAPE.

A Keith Martin, National Research Institute for Oceanology,
CSIR, P O Box 320, STELLENBOSCH, 7600 CAPE.

Tim P McClurg, National Institute for Water Research,
CSIR, P O Box 17001, CONGELLA, 4013 NATAL.

Michael H Schleyer, Oceanographic Research Institute,
P O Box 10712, MARINE PARADE, 4056 NATAL.

Eckart H Schumann, Department of Oceanography, Univ. of Port Elizabeth,
P O Box 1600, PORT ELIZABETH, 6000 CAPE.

Geoff Toms, National Research Institute for Oceanology,
CSIR, P O Box 320, STELLENBOSCH, 7600 CAPE.

Rudy van der Elst, Oceanographic Research Institute,
P O Box 10712, MARINE PARADE, 4056 NATAL.

CONTENTS

Chapter 1. INTRODUCTION
E H Schumann 1

Chapter 2. PHYSIOGRAPHY, STRUCTURE AND GEOLOGICAL EVOLUTION OF THE NATAL CONTINENTAL SHELF
A K Martin and B W Flemming 11

Chapter 3. SEDIMENT DISTRIBUTION AND DYNAMICS ON THE NATAL CONTINENTAL SHELF
B W Flemming and E R Hay 47

Chapter 4. CLIMATE AND WEATHER OFF NATAL
I T Hunter 81

Chapter 5. PHYSICAL OCEANOGRAPHY OFF NATAL
E H Schumann 101

Chapter 6. INORGANIC NUTRIENTS IN NATAL CONTINENTAL SHELF WATERS
R A Carter and J d'Aubrey 131

Chapter 7. PLANKTON DISTRIBUTIONS IN NATAL COASTAL WATERS
R A Carter and M H Schleyer 152

Chapter 8. BENTHOS OF THE NATAL CONTINENTAL SHELF
T P McClurg 178

Chapter 9. SHELF ICHTHYOFAUNA OF NATAL
R van der Elst 209

Chapter 10. POLLUTION AND EFFLUENT DISPOSAL OFF NATAL
A D Connell 226

Chapter 11. THE RICHARDS BAY MARINE DISPOSAL PIPELINES
D A Lord, G Toms and A D Connell 252

Appendix. THE R V MEIRING NAUDÉ 270

Chapter 1

INTRODUCTION

Eckart H Schumann
Department of Oceanography
University of Port Elizabeth

The ocean areas surrounding the world's land masses have always been important in man's social and economic structures. They are a source of food and a place for recreation, and provide vital transport routes. More recently valuable minerals have been procured; on the other hand, the ocean has regrettably also been regarded as a convenient sink for man's waste products.

In order to utilize these coastal areas fully, it has become more and more essential to understand the processes operating there. This knowledge must extend over a variety of disciplines, including studies of rock and sand strata, the atmosphere, the ocean itself and all the chemical interactions and biota which depend on that environment. In this way overall perspectives can be achieved, enabling better management decisions to be made.

This monograph does not seek to cover the whole offshore area. The "coastal ocean" referred to here lies beyond the surf zone but does not extend into the deep ocean. Essentially, it will comprise the relatively narrow continental shelf off Natal, but because no area such as this can be treated in isolation, it will be discussed in relation to any other factors which may be important in obtaining an understanding of the region. In particular, the influences of the deep ocean, land morphology and wider weather patterns of the area will be discussed.

This introductory chapter is intended to set the scene by giving historical and regional perspectives, and also a background to scientific endeavour in the region. The contributions of the various chapters will be assessed and it should then also become clear why publication of this monograph is opportune at this stage.

HISTORICAL AND REGIONAL PERSPECTIVES

Natal forms one of the four provinces of the Republic of South Africa, the others being the Cape, Transvaal and Orange Free State. In land area it is the smallest of the provinces, with an eastern seaboard of about 570 km washed by the south-west Indian Ocean.

In comparison with most other countries, Natal has a relatively short recorded history (Brooks and Webb, 1965). The name 'Natal' was given to it by Vasco da Gama, from whose ships land was sighted on Christmas Day, 1497, on their epoch-making journey from Portugal to India. However, the traders passed by this coast, and for some years the only Europeans to go ashore were those who had been shipwrecked; they found, in general, that the indigenous Nguni people were hospitable and friendly.

The 'Port of Natal' was recognised as the safest anchorage on the coast and was used mainly for replenishment of water supplies and in times of distress. Nonetheless, it was not until the 1820s that English settlers first established a permanent presence in the area. In 1835 a town was founded and named Durban, in honour of the then Governor of the Cape.

Subsequent developments over the following half century saw the interior opened up to further European settlement. Earlier, the Nguni in the region had been unified by Shaka, chief of the Zulu tribe, and thereafter also became known as Zulus. As could be expected, clashes occurred over control of the land, not only between the Zulus and the newcomers, but also between different groups of settlers.

It was not until 1903, when the military might of the Zulus had been broken, and the Transvaal and Orange Free State Republics had succumbed to the British forces, that the boundaries of Natal were finally fixed. These boundaries also defined the Province of Natal as incorporated into the Union of South Africa in 1910. The present situation is depicted in Figure 1.1.

There are 73 significant rivers and estuaries along the Natal coast with outlets to the ocean (Begg, 1978). With the Drakensberg mountains reaching heights of over 3 000 m along the western boundary of the province, the catchment areas are relatively small. Nonetheless, good rains over much of the region - over 1 000 mm per

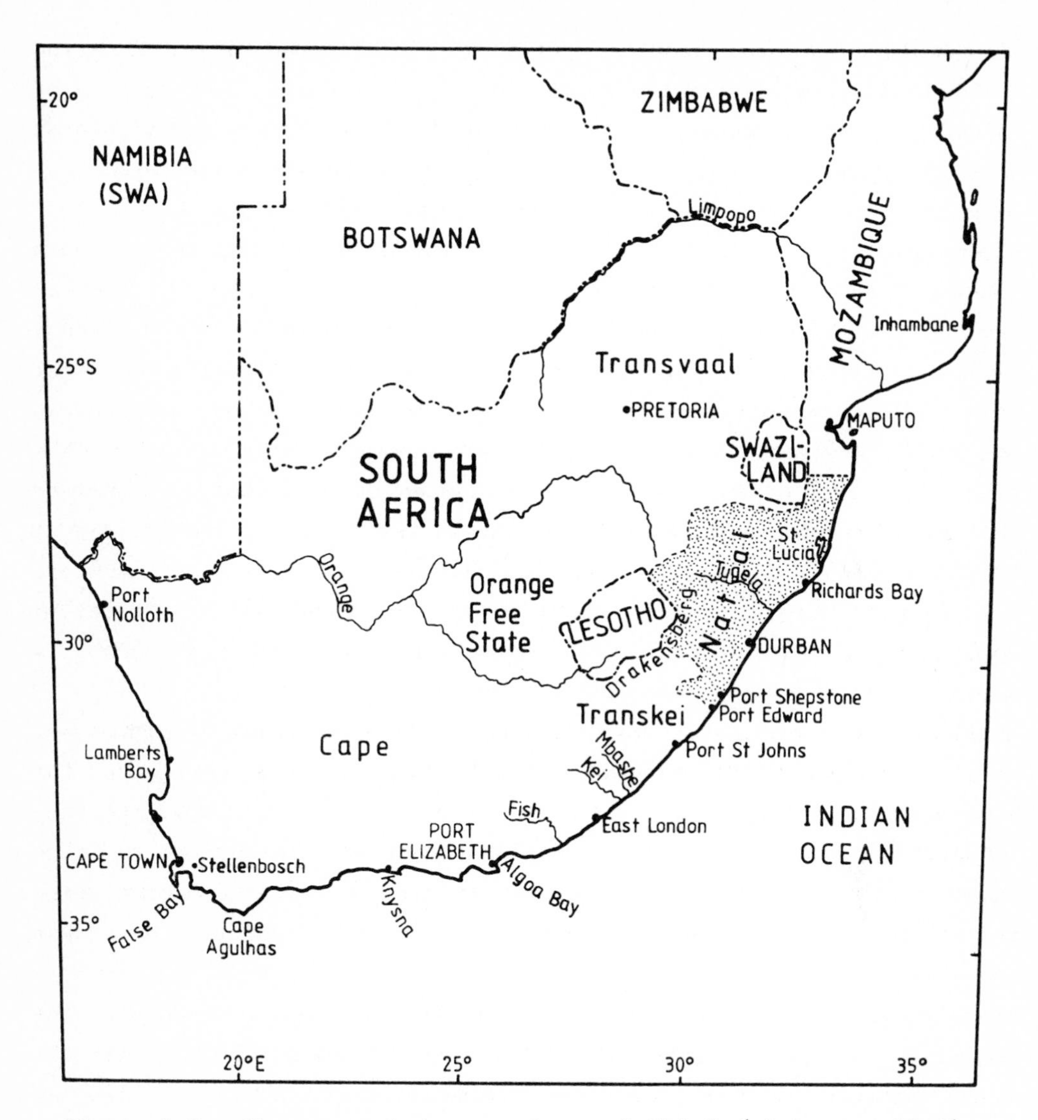

Figure 1.1 The present-day province of Natal (shown speckled), within the political sub-division of the African sub-continent. Names of the main towns and geographic features are given as well as other points mentioned in subsequent chapters.

annum in places - have led to Natal being termed the 'garden province', and the Tugela river ranks second in South Africa in terms of volume outflow. However, most of the rain falls in summer, and during the dry winter months the mouths of most estuaries are closed by sandbars.

The northeastern subdivision of Natal is called Maputaland, and forms the southern section of a coastal plain that extends northwards to Somali. At the southern point of this coastal plain lies the largest estuarine system in Africa, namely, St Lucia, with an area of roughly 300 km^2. There is evidence that Arab traders reached this area during the period about 1250 to 1350, but until today Maputaland has remained relatively undeveloped (Bruton and Cooper, 1980).

Further south, Durban underwent rapid development, and is now South Africa's foremost port in terms of volume of general cargo: in 1981/82 over 21.5 million tons were handled (South Africa, 1984). Other towns, such as Port Shepstone and Port Edward, bear witness to the fact that some of the larger estuaries were used as ports by early European settlers. However, none of these operate as such today.

All the same, industrial development made it necessary that another harbour be developed further north, and in 1972 construction started at Richards Bay. The harbour was commissioned in 1976, primarily to handle bulk cargoes, especially coal. The multi-purpose facility handled 31.3 million tons of cargo in 1981/82, a total which is expected to increase to 44 million tons later in the decade.

Durban, and the Natal coastal areas to the north and south, are tourist Meccas, particularly for holidaymakers from the industrial interior of South Africa. The main attraction is undoubtedly the warm water of the Indian Ocean, combined with wide stretches of fine beaches. Amenities are therefore geared to cater for this demand, with activities such as sport fishing and boating forming important components.

It is therefore clear that the sea plays an important part in the social and economic structure of Natal. The coastal areas, with all their attractions and resources, create the environment for an important source of revenue. The two ports, with all their associated infrastructure, are essential in the whole economic pattern of South Africa.

SCIENTIFIC ENDEAVOURS

With the acknowledged importance of the sea, it is surprising to find only a few organisations doing a limited amount of oceanographic work off Natal. This is in contrast to the situation in the Cape, where in the Cape Town area alone there are four or five organisations active in the marine field. Even though the Natal Government commissioned a survey of the fishing grounds in 1900 to 1901, and these surveys continued into the 1920s, it did not lead to the establishment of a marine institute. This eventually was left to a group of private individuals, from an idea voiced around a campfire in 1947. As a consequence, the South African Association for Marine Biological Research was formed in 1952, with the Aquarium and Oceanographic Research Institute (ORI) being opened in Durban in 1959 (SAAMBR, 1960). Funds were obtained from commercial and other organizations, including the City Council, while all the profits from the aquarium were devoted to research. Close ties were established with the University of Natal, although there has never been a university department established there specifically for marine research. Collaboration and emphasis have been largely dependent on individual interests.

Initial research at ORI was aimed at combating the shark menace to bathers off Natal's beaches, and this led to the establishment, by Provincial statute, of the Natal Anti-Shark Measures Board in the late 1960s. This organization is responsible for erecting and maintaining the nets that have rendered the Natal beaches safer in terms of shark attacks on bathers for almost two decades.

More recent research at ORI includes studies of marine and estuarine fish, prawns, rock lobsters, mussels, turtles, marine productivity and ecology. However, the lack of an adequate research vessel has meant that the activities have been primarily concentrated in the nearshore regime.

In about 1960 the South African Council for Scientific and Industrial Research (CSIR) also initiated oceanographic work off Natal. This came about as a result of effluents being disposed of to sea, and the recognition that "coastal waters do not have an infinite capacity for absorbing foreign matter" (CSIR, 1964). More information was required on the structure and dynamics of the ocean environment in order that sound advice could be given for the disposal of such effluent. Three institutes were involved, namely the National Institute for Water Research (NIWR), the National Physical Research Laboratory (NPRL) and the National Mechanical Engineering Research Institute (NMERI), all parts of the CSIR.

Investigations involved the use of tracked floats, drift cards, moored indicating buoys, diffusion studies using dyes and models, bacterial tracers and aerial photography, while monitoring equipment was also developed. Where possible, ships such as whalers and an ex-Navy crash boat were used. As the expertise developed, work was also done for a few commercial enterprises.

The capabilities of the group improved dramaticaly in 1968, when a research ship, the RV Meiring Naudé, was commissioned for the Physical Oceanography Division of the NPRL (see Appendix). This now meant that regular investigations further offshore could be commenced, including hydrographic surveys of the whole south-east coast of Southern Africa. The development of Richards Bay as a harbour meant that a number of measurement cruises were carried out there in the early 1970s. Closer inshore, the Hydraulics Research Unit of the NMERI was involved in harbour and breakwater design, while biological and chemical monitoring was undertaken by the NIWR.

In 1974 the National Research Institute for Oceanology (NRIO) was established, incorporating both the Physical Oceanography Division and the Hydraulics Research Unit. A decision was taken at that time to transfer the latter to the NRIO headquarters in Stellenbosch, near Cape Town, which meant that only contract work has been done in coastal engineering aspects since then.

There was also a change in emphasis of the investigations in the ocean areas off Natal. More basic research was carried out, biological and chemical work expanded, and detailed geological and sediment surveys initiated.

However, funding was not maintained at an adequate level to keep the NRIO groups in Durban viable, and in about 1980 a decision was made to transfer these to Stellenbosch as well. Nonetheless the RV Meiring Naudé has been kept in Durban, to be used by research groups in a variety of disciplines. This has meant that most of the physical and plankton work on the shelf areas has ceased although geological investigations have continued at a lower key; at present the NIWR and ORI are the main organizations working there. Occasionally groups from museums use the ship, with the emphasis of their work on taxonomy.

DISCUSSION

It is clear from the above description that a fair amount of data on the coastal ocean off Natal has been collected. Much of it has been analysed and published; however, it has not yet been presented in any collected form. The object of this volume is the synthesis of the information that is available. Moreover, there seems to be little prospect of any immediate major upturn in oceanographic activity off Natal. Such a volume as this is therefore unlikely to date quickly, and should provide a statement of the knowledge of the area for some time in the future. It should also prove to be of value to any organization or individuals requiring information on a variety of marine subjects.

The actual specification of the coastal ocean off Natal as the subject for this volume has been done as an expediency to embrace a political entity. There are naturally some areas which have been more thoroughly investigated, particularly near the main urban areas; much less work has been done off Maputaland. Moreover, environmental and ecological factors are certainly not bound by any arbitrary political subdivisions.

An attempt has been made to cover as many branches of the marine sciences as possible. The interdependence of these various disciplines in investigating the complex coastal ocean region off Natal is apparent, but the approach adopted here is not so much interdisciplinary as multidisciplinary. In part this is because of the way in which the various research programmes developed within different organizations, but it is also because of the difficulty of initiating and carrying through truly interdisciplinary research over a broad front.

Nonetheless, it is the physical sciences which are the more independent, and it is the environment which sets the conditions in which the marine biota need to find a niche in which to live. The most obvious physical component is the base structure, namely the geology, and a description is given of the physiography and geological evolution of the region. The narrow shelf, with its steep slope going down to the Natal valley, identifies important limiting conditions which affect many of the shorter-term oceanographic processes occurring off Natal.

By the same token, the gentle Natal bight north of Durban provides a slightly wider shelf with identifiably different conditions.

Various regimes are identified in the more recent, and ongoing, sedimentary processes. These are closely associated with the oceanic circulation patterns, and the strong currents are shown to produce unique, mobile sediment structures; biologically these have vital consequences in the resulting unstable substratum.

The most important oceanographic feature is undoubtedly the Agulhas Current, one of the major western boundary currents of the world's oceans. The core of the Current generally lies offshore of the shelf break, but its influence can be felt over much of the shelf itself. Apart from the physical effects, the Agulhas Current also transports tropical and subtropical species southwards, providing an additional supply of larvae. In the ichthyofauna, close links are established with the rich Indo-Pacific region.

The climate of the Natal coastal belt is described as humid subtropical, with a warm summer. It lies within the southern sub-tropical high pressure belt, coming strongly under the influence of eastward migrating high pressure systems. There are marked seasonal variations, but coastwise-parallel winds dominate at all times. An important aspect is the so-called "coastal-low", migrating northwards and being sometimes associated with strong southwesterly "busters".

These winds play an influential part in the dynamics of the coastal ocean. On the wider shelf the currents are predominantly wind-driven, resulting in fluctuating regimes with long residence times for water masses; clearly this could be vital for larvae in the area. During summer heavy rains can cause large quantities of sediment to be discharged to sea by the rivers, with adverse effects on certain categories of benthos. This has increased in recent times because of the severe erosional consequences of agricultural malpractices; evidence for this can be found in the increased sediment accumulation rates in the Natal Valley. On the shelf itself, the existence of well-defined mud depocentres indicate that those areas are frequently occupied by closed eddy systems; at present there are not sufficient current measurements to confirm such conclusions.

Inorganic nutrients form the basis of the food chain. The Agulhas Current itself has low concentrations of such nutrients, while on the shelf there are positive gradients with depth. These nutrients can be brought nearer the surface by sporadic, small-scale upwelling events, resulting in patchy distributions. Correspondingly, phytoplankton production rates increase although the biomass remains generally low. On the other hand zooplankton biomass reaches high levels in the winter/spring period, being comprised mainly of copepods.

There is also a marked seasonality in the fish populations. The summer fish tend to be wide-ranging, while most of the winter species are either endemic or comprise isolated populations in South African waters. When the waters cool, migration occurs from the temperate Cape waters, with the most dramatic being the annual 'sardine run'. This represents a significant input of nutrients to the Natal ecosystem.

The recent effects of man's involvement have had major implications for the Natal coastal areas. Housing and industrial developments have brought with them the concomitant problems of waste disposal, and the easy option - that of dumping to sea - has been used in many cases; in the past this resulted in appreciable levels of pollution in certain areas. However, the establishment of water quality criteria, and the enforcement of such regulations, has meant that there are now generally few pollution problems off Natal. To a large measure this can also be ascribed to the dynamic physical environment, in which potential pollutants are rapidly dispersed and removed.

The 1970s saw the construction of the harbour at Richards Bay. This was a unique development, and one of the pleasing features was the attempt to preserve a part of the original bay as a sanctuary area; this involved the construction of a berm wall across the bay, and then opening a new mouth through the coastal dunes (Begg, 1978). Two outfall pipes 4 and 5 km long have also been constructed to discharge effluents into the coastal ocean. Since such developments require a knowledge of the environment and the effects of the effluents on the biota, and that moreover coastal engineering has become an important influence in the area, a special chapter is devoted to this subject.

A worrying and immediate problem concerns the exploitation of Natal fish. The catch per unit effort has increased dramatically, while valued species have all but disappeared; that these overfished species are all endemic demonstrates their vulnerability to this excessive exploitation. While the total tonnage caught is small, the value of the fishery lies in its recreational and tourist potential, and it is therefore important that it remains healthy.

In conclusion, it can be stated that a reasonable basis has been laid for understanding many of the physical, chemical and biological processes in the coastal ocean off Natal. However, much work remains to be done to clarify details, and on the biological side many species still await discovery. Furthermore, the interdisciplinary nature of the area needs to be addressed, in order to gain an understanding of the functioning of the whole ecosystem.

REFERENCES

BEGG, G (1978). **The estuaries of Natal.** Natal Town and Regional Planning Report Vol. 41, 657 pp.

BROOKS, E H and C DE B WEBB (1965). **A history of Natal.** University of Natal Press, Pietermaritzburg, South Africa, 371 pp.

BRUTON, M N and K H COOPER (eds.) (1980). **Studies on the ecology of Maputaland.** Rhodes University and the Natal Branch of the Wildlife Society of Southern Africa, 560 pp.

CSIR (1964). **The marine disposal of effluents with particular reference to the Natal coast,** 9 pp.

SOUTH AFRICA 1984. Official yearbook of the Republic of South Africa. Tenth edition - 1984. Compiled and edited by the Department of Foreign Affairs and Information. Chris van Rensburg Publications, Johannesburg, 1058 pp.

SOUTH AFRICAN ASSOCIATION FOR MARINE BIOLOGICAL RESEARCH. Bulletin No. 1. December 1960, 26 pp.

Chapter 2

PHYSIOGRAPHY, STRUCTURE AND GEOLOGICAL EVOLUTION OF THE NATAL CONTINENTAL SHELF

A K Martin
National Research Institute for Oceanology, CSIR

B W Flemming
Senckenberg Institut, Wilhelmshaven, West Germany

INTRODUCTION

Prior to the Jurassic Period, 180-135 million years ago (Ma), Southern Africa, and in particular Natal, lay in a central position within the ancient super-continent, Gondwanaland (Du Toit, 1937; Barron _et al_. 1978; Martin and Hartnady, 1986). The history of the Natal continental margin therefore begins with the break-up of Gondwanaland in the Jurassic and Early Cretaceous periods. The continental margin, which formed the boundary of the "new" African continent, evolved as the Indian Ocean expanded during the dispersal of the Gondwanide fragments. As new drainage patterns were established in response to rifting and formation of new ocean basins, geological formations which pre-dated break-up were masked by Late-Jurassic to Recent sediments. The nature of the continental shelf and margin was controlled by the type of rifting, drainage patterns, terrestrial and marine sediment supply, sediment loading and subsidence as well as the currents, chemistry and sea-level of the adjacent ocean. Models of these processes have been provided at a global scale by Hay _et al_. (1981), and the sedimentary basins around Africa are discussed by Dingle (1982).

The present coastal plain and continental shelf represent emerged and submerged portions of a continuous feature, the separation between the two parts at any time depending on the relative sea-level. Geological endeavour in the Natal coastal region has a relatively long history (for example, Anderson, 1907; McCarthy, 1967; Frankel, 1972; Flores, 1973; Forster, 1975; Kennedy and Klinger, 1975; Siesser and Miles, 1979). Offshore investigations are more recent, being initiated by the

Geological Survey, University of Cape Town group (Moir, 1974; Dingle and Siesser, 1977; Dingle *et al.* 1978; Martin, 1981) and continued by the National Research Institute for Oceanology (NRIO) (Flemming, 1981; Martin, 1985; Martin and Flemming, 1986). Unlike most of the oceanographic work on the shelf (Chapter 5) geological work is continuing and much material has yet to be evaluated. Some commercial exploration data have been released (Du Toit and Leith, 1974; McLachlan and McMillan, 1979) and these have been incorporated into a comprehensive review of the Mesozoic and Tertiary geology of southern Africa (Dingle *et al.* 1983). More information on pre-break-up geology has been published by Hobday (1982) and Tankard *et al.* (1982).

The purpose of this chapter is to describe the geology of the Natal continental margin, with emphasis on how geological processes and features affect the Natal coastal ocean. The morphology of the shelf and continental slope is described in relation to the adjacent ocean basin, the Natal Valley (see Figure 2.1). Beginning with continental break-up, the evolution of the area is outlined by considering the structure and sedimentology of successively younger stratigraphic units of the Mesozoic and Cenozoic Eras.

PHYSIOGRAPHY

The Natal continental margin descends into the adjacent ocean basin, the Natal Valley (Figure 2.1). To the east of this lies the aseismic Mozambique Ridge which separates the Natal Valley from the Mozambique Basin. The northernmost Natal Valley as far south as 29°S and the Tugela Cone form shallow (<2 000 m) marginal plateaux extending from the continental margin. More abyssal depths (~3 000 m) extend seaward of these features, where the Natal valley deepens southwards towards the Transkei Basin. Thus, although the region is a passive continental margin, having been formed by rifting rather than continental collision, it does not display typical passive margin morphology.

Two contrasting morphological types are displayed by the Natal continental margin (Figure 2.2). North of 28°30'S, and again south of 30°20'S, the continental shelf is very narrow and the continental slope is steep, whereas between 28°30'S and 30°20'S, the shelf is wide and

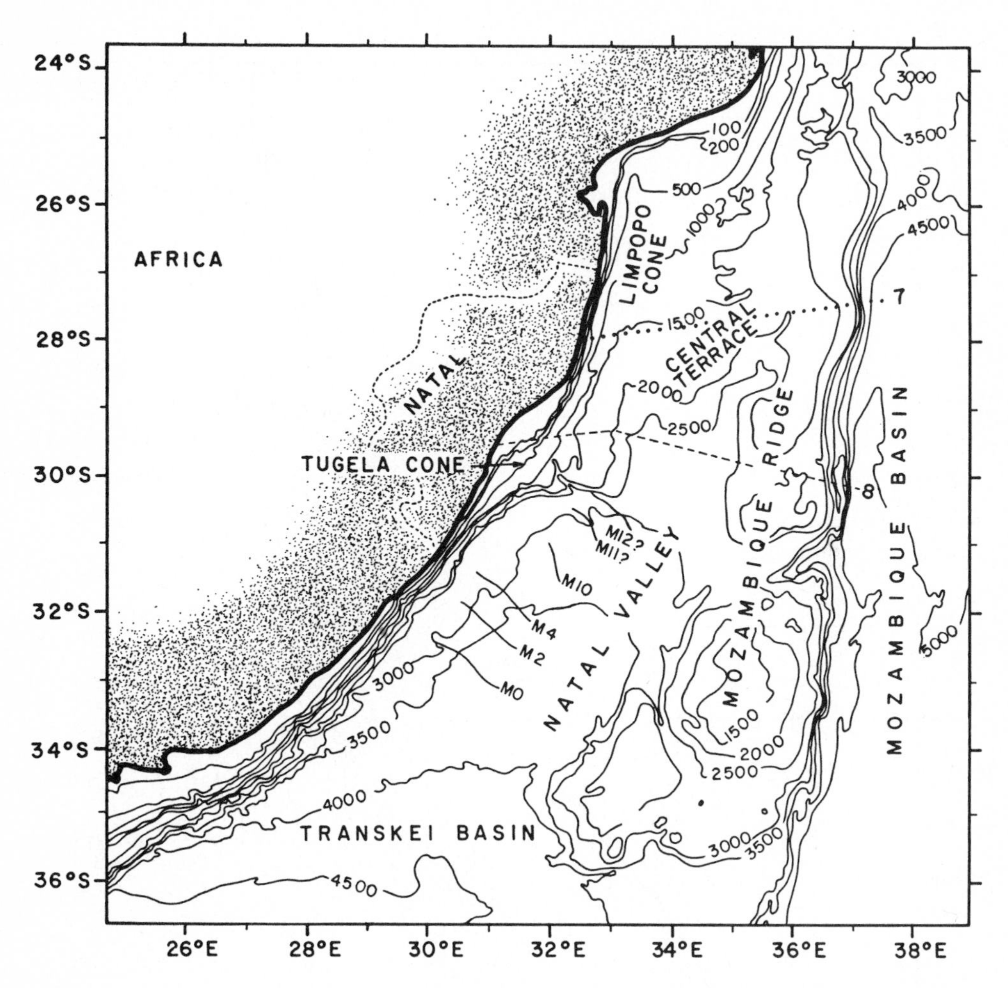

Figure 2.1 Physiography of the south-west Indian Ocean, (after Martin, (1981) and General Bathymetric Chart of the Oceans (GEBCO) sheet 509, Canadian Hydrographic Service, Ottowa). Numbered solid lines (example M10) represent seafloor spreading magnetic anomalies (after Martin et al. 1982). Dotted lines 7 and 8 are cross-sections shown in Figures 2.7 and 2.8.

the continental slope is gentle (Table 2.1). The broader shelf area is an expression of the Tugela Cone, a triangular-shaped marginal plateau extending 220 km south-east of the Tugela River (Goodlad, 1986). This feature displays varied topography, comprising terraces, hummocks, peaks, valleys and scarps and is cut by two large canyons - the Tugela canyon and one at 29°30'S. The steep sections of the continental slope are also cut by deep canyons: five major canyons cut the Zululand

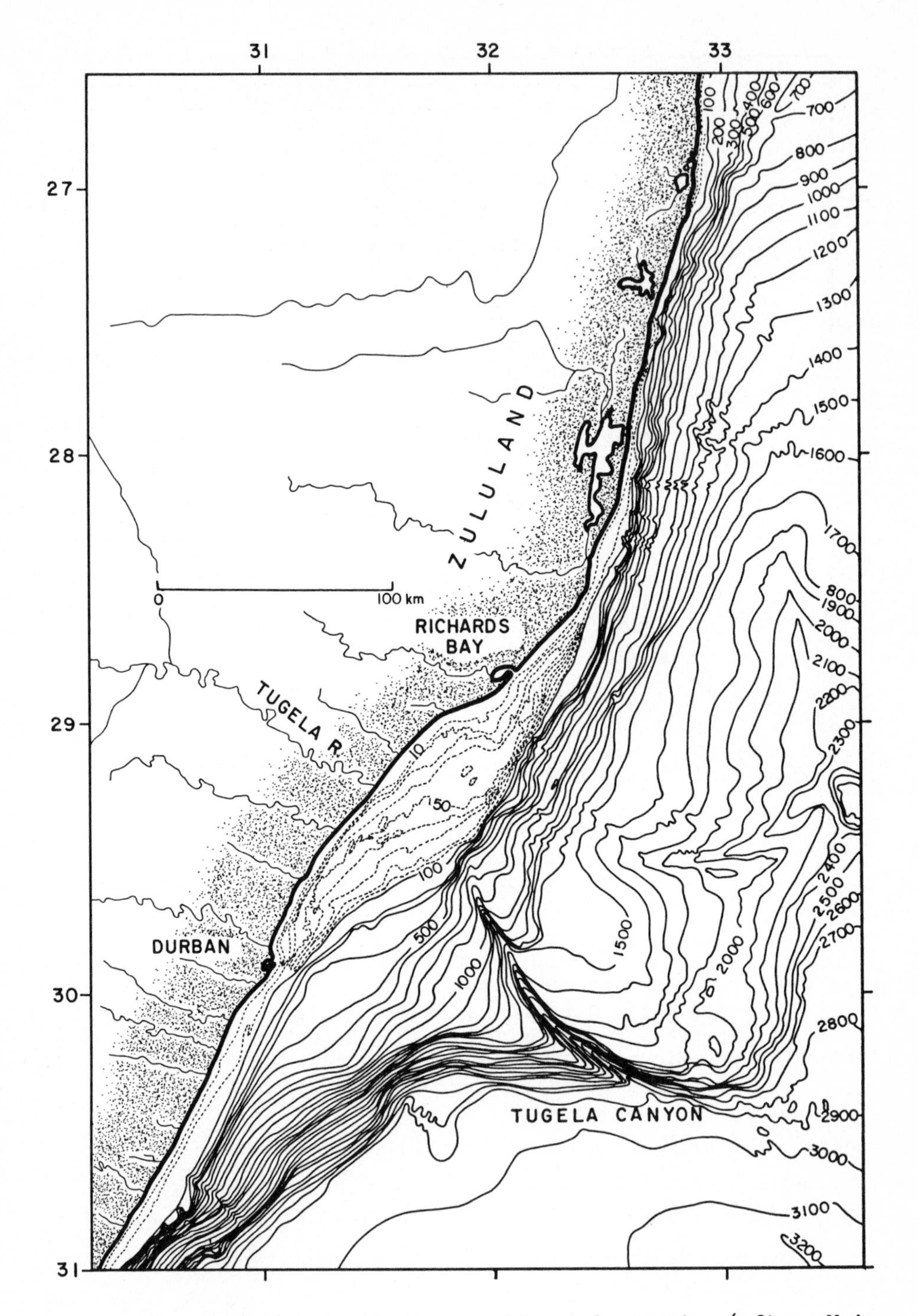

Figure 2.2 Bathymetry of the continental margin (after Moir, 1974; Goodlad, 1979; Martin, 1984). More detailed bathymetry of parts of the shelf is given by Birch (1981) and Hay (1984).

Table 2.1 Bathymetric Data

CONTINENTAL SHELF			CONTINENTAL SLOPE		
Area	Width (km)	Shelf Break Depth (m)	Max. Relief (m)	Average Slope	Max. Slope
North of 28°30'S	2-7	45-112	800-1500	1:21 (2.9°)	1:10 (5.7°)
28°30'S-30°20'S	Up to 45	100-112	2400-2900	1:60 (1°)	1:17 (3.37°)
South of 30°20'S	10	80-90	2900	1:14 (4.1°)	1:35 (16°)
World Average	74	132	Av.4000	1:13 (4.3°)	

continental slope north of 28°30'S and at least five dissect the area south of 30°20'S (Martin, 1981; Birch, 1981).

The inner continental shelf is smooth because of its cover of Holocene sediments (Birch, 1981; Hay, 1984), whereas the mid- and outer-shelf is punctuated by a series of rugged linear shoals. These structures (Figures 2.13 and 2.14) form the major topographical features of the shelf and affect Holocene sedimentation (Chapter 3), and may influence the flow of the Agulhas Current on the shelf (Hay, 1984).

CONTINENTAL BREAK-UP: INITIATION OF THE CONTINENTAL MARGIN

Because sediments in the Karoo Supergroup were derived from east of the present coastline, it has long been recognized that a land-mass lay to the east of Natal in late Palaeozoic/Early Mesozoic times (Du Toit, 1937). The only seafloor spreading magnetic anomalies identified off Natal extend south-westwards from the Tugela Cone (Martin et al., 1982, and Figure 2.1). These indicate that a short seafloor spreading ridge initiated around 132 Ma between the southern face of the Tugela Cone and the Falkland Plateau, a part of the South American plate (Figure 2.3). During initial movement, the Falkland Plateau slid southwestward past the southern Natal continental margin along a long transform fault. The spreading ridge associated with this movement migrated south-west of southernmost Natal by 127 Ma. A new reconstruction (Martin and Hartnady, 1986), which uses all known magnetic anomalies and fracture zones as constraints, suggests that East Antartica fitted against Mozambique and the Lebombo Mountains (Figure 2.3). Antarctica

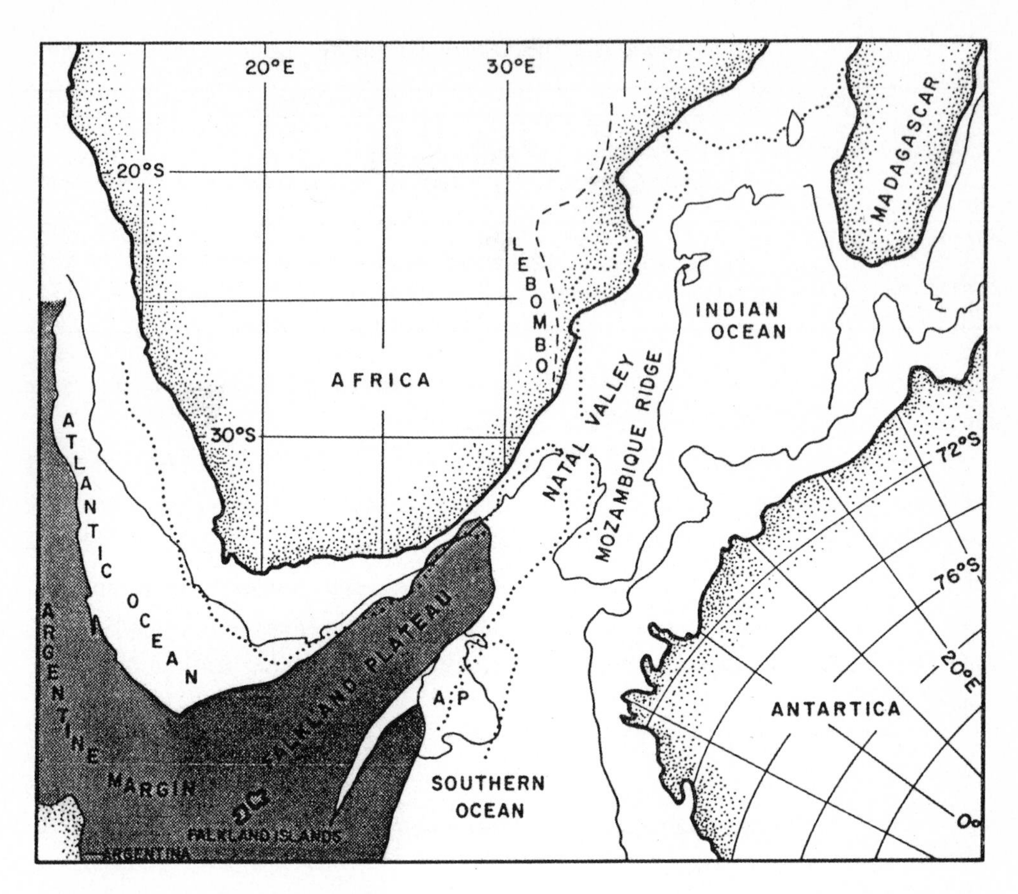

Figure 2.3 A reconstruction of Africa, Madagascar, Antarctica and South America as they were at the time of sea-floor spreading magnetic anomaly M2 (123 Ma); each continent is outlined by the 3 000 m isobath. Note the incipient ocean basins. Dotted lines mark the 3 000 m isobaths of South America, the Falkland Plateau and East Antarctica in their pre-break-up positions, that is prior to 153 Ma (after Martin and Hartnady, 1986). A.P. = Agulhas Plateau.

began moving south-south-eastward relative to Africa before 153 Ma, sliding past Africa along a transform fault east of, and parallel to, the Lebombo mountains.

Oceanic crust, emplaced against continental crust at transform faults, cools and subsides becoming uncoupled from the adjacent continental crust. This is the underlying cause of steepness along sheared continental margins (Scrutton, 1979). In contrast, at rifted margins, heating is followed by thinning through listric faulting, subsidence and final break-through of oceanic crust (LePichon and

Sibuet, 1981; Scrutton, 1982).

The Natal continental margin therefore comprises three separate zones, corresponding with the physiographic regions outlined earlier. A sheared (transform-faulted) margin developed east of the Lebombo Mountains and along the eastern face of the Tugela Cone 153-132 Ma. Some movement may have occurred prior to 153 Ma during the earlier stages of Lebombo vulcanism. Between 135 and 132 Ma seafloor spreading began along a short rifted section on the southern Tugela Cone. Simultaneously a second sheared margin began to develop south of 30°20'S on the Agulhas margin.

In this model for the break-up of Gondwanaland (Figure 2.3), volcanic rocks underlying coastal Mozambique, the Central Terrace and the Mozambique Ridge are regarded as oceanic, having been formed during the early stages of seafloor spreading (Martin and Hartnady, 1986). If Africa is moved relative to the hot-spot reference frame (Duncan, 1984), the Bouvet hot-spot is placed close to the incipient rifts in the Natal Valley, and may have led to the thickened oceanic crust in a similar way to the construction of thick oceanic crust under Iceland today.

STRATIGRAPHY

The geological map (Figure 2.4) shows that pre-break-up rocks include the Archean Kaapvaal Craton, the mid-Proterozoic Namaqua-Natal gneissic terrane, sandstones of the Palaeozoic Natal Group and Late Carboniferous to Early Jurassic sandstones and shales of the Karoo Supergroup. Uppermost Karoo volcanic rocks of the Lebombo and Drakensberg Mountains mark the initiation of the continental margin (Figure 2.5).

Immediately overlying the volcanic rocks in Mozambique and northern Natal is a basal continental conglomerate and the Sena sandstone (Flores, 1973; McLachlan and McMillan, 1979). Although break-up began at least by 153 Ma (Simpson *et al*. 1979; Martin and Hartnady, 1986) the earliest marine sediments are Barremian in age (Forster, 1975; McLachlan and McMillan, 1979), whereas at the southern end of the Lebombo, successively younger formations overlie volcanic rocks

(Kennedy and Klinger, 1975). The Barremian to Aptian sequence, the Makatini Formation, is developed only in Zululand, whereas on the shelf north of Durban the oldest Cretaceous rocks date from the Cenomanian/Turonian and overlie the Dwyka Formation of the Karoo Supergroup (Figure 2.5, compare columns for Zululand and the JC-1 borehole). The Albian-Cenomanian Mzinene Formation overlies an Aptian-Albian boundary sequence. The unconformity above the Mzinene Formation occurs throughout the area extending offshore to the Natal Valley where it is seen in seismic profiles as a prominent unconformity called McDuff (Dingle et al. 1978). It has been correlated with a reflector abutting basement near the JC-1 borehole, and with an unconformity on the Mozambique Ridge (Simpson, Schlich et al. 1974), although the hiatus does not occur in south-eastern Mozambique (Flores, 1973). On the Mozambique Ridge this hiatus marks a fundamental change from reducing to oxidizing conditions. A similar chemical change occurred later in Mozambique in the Turonian, but may represent a later stage of one progressive event (Girdley et al. 1974).

Figure 2.4 Geological map. Onshore geology is simplified from the geological map of the area (Department of Mines and Energy, Pretoria, 1984). Offshore geology is an update of the map of Dingle and Siesser (1977), and is based on NRIO seismic profiles correlated to available dated samples and the JC1 borehole (Du Toit and Leith, 1974; Siesser, 1977). Key: 1) Archaean (3 billion years old) granites and granodiorites, and the overlying meta-sediments of the Pongolan Sequence; 2) mid-Proterozoic (1 000 Ma) gneisses of the Mapumulo and Tugela Groups, Natal Province of Namaqua-Natal Metamorphic Terrane; 3) sandstones of the Palaeozoic Natal Group; 4) sandstones and shales of the Late Palaeozoic Early Mesozoic Karoo Sequence; 5) Late Karoo (Jurassic) basalts and rhyolites of the Lebombo Group; 6) Early Cretaceous conglomerates, sandstones and siltstones of the Makatini and Mzinene Formations; 7) Late Cretaceous marine sandstones and siltstones of the St Lucia and Umzamba Formations; 8) pre-Oligocene Tertiary strata; 9) Miocene strata; 10) Pliocene strata offshore and Late Miocene - Early Pliocene Pecten Bed onshore; 11) slump facies affecting strata as recent as post Miocene-Pliocene boundary; 12) post-slump prograding shelf-edge sequence; 13) Pleistocene - Recent sands; note that, offshore, Pleistocene aeolianites (Figure 2.14) and Holocene unconsolidated sediments (Chapter 3) have been omitted for clarity; 14) slump glide-plane scar; 15) canyons; 16) faults. Dotted lines numbered 6, 9, 10, 11 and 12 are locations of Figures 2.6 and 2.9-2.12.

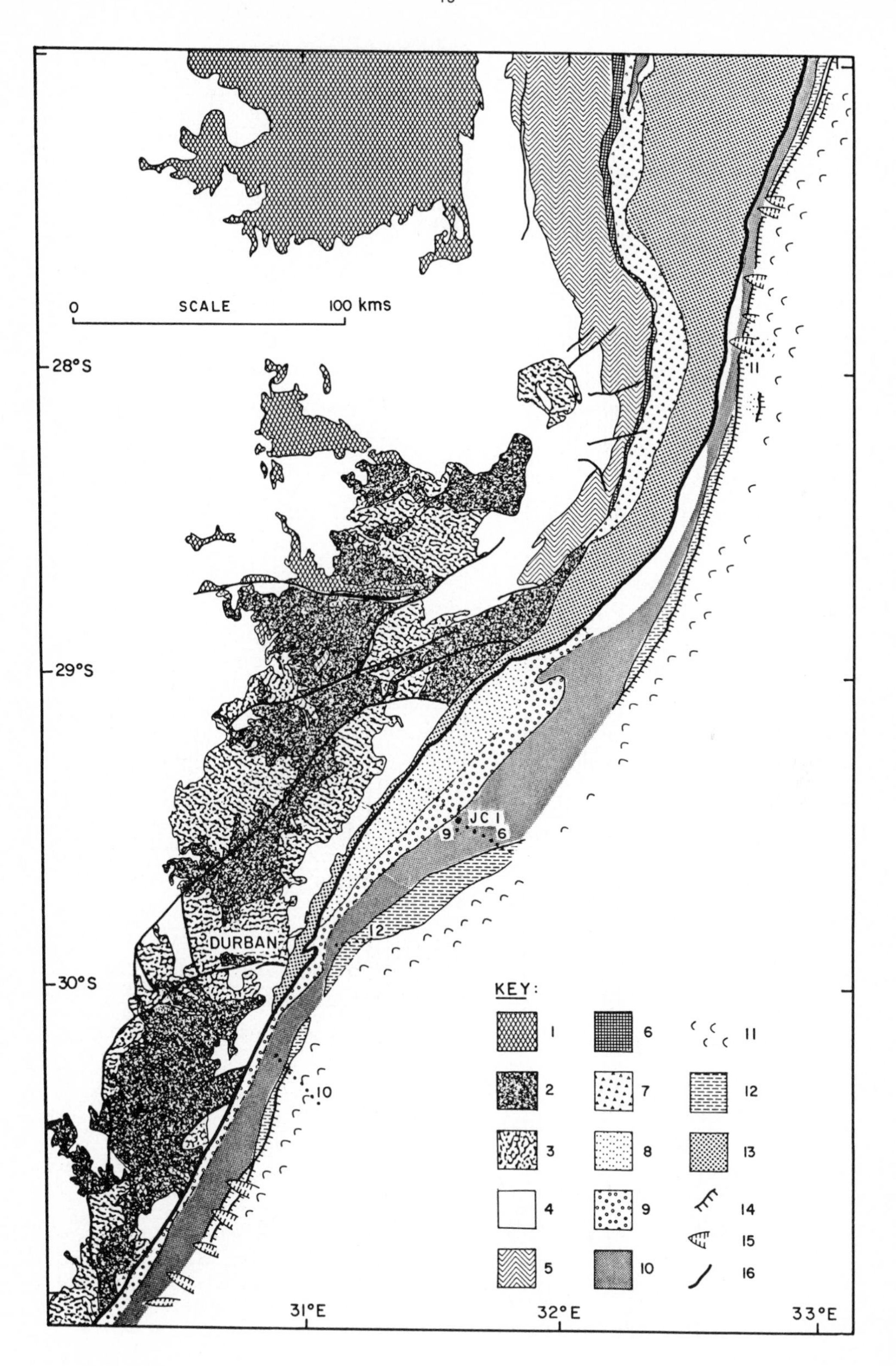
0
SCALE
100 kms
28°S
29°S
30°S
31°E
32°E
33°E
JC 1
9
6
12
11
10
DURBAN
KEY:
1
2
3
4
5
6
7
8
9
10
11
12
13
14
15
16

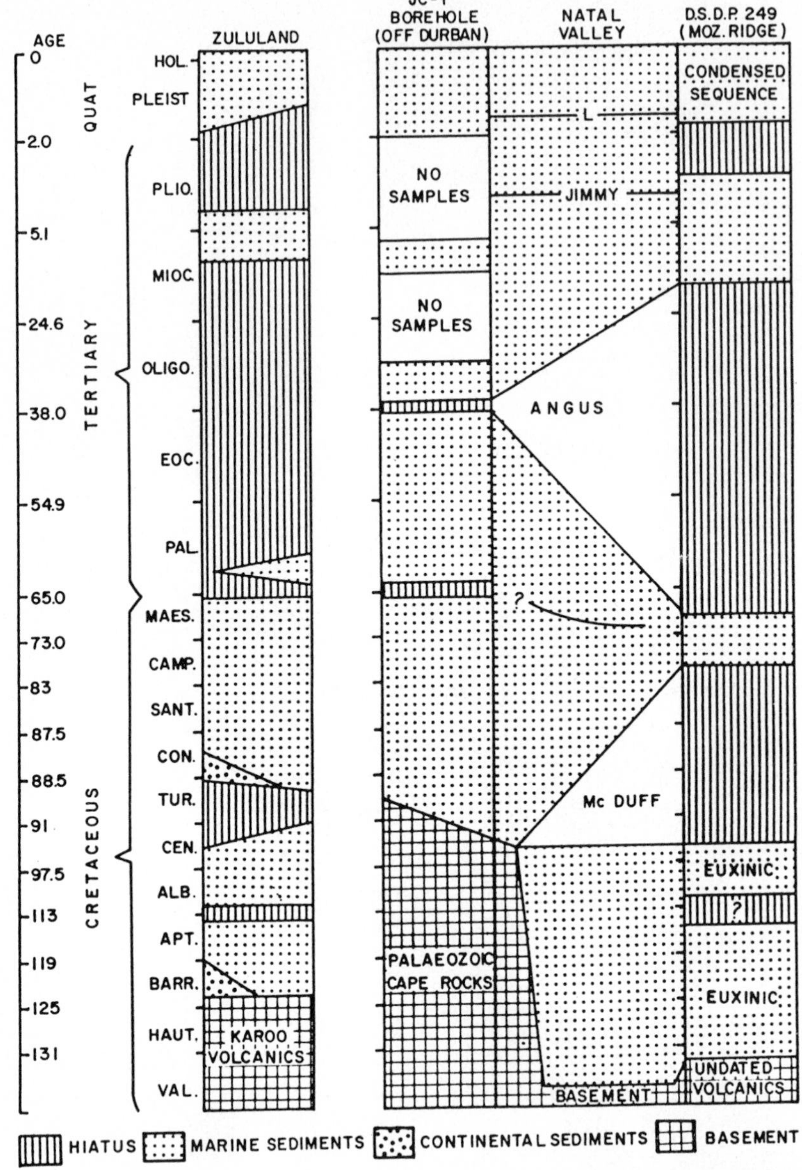

Figure 2.5 Stratigraphic columns for the Natal continental margin after Du Toit and Leith (1974), Simpson, Shlich _et al_. (1974), McLachlan and McMillan (1979), and Martin _et al_. (1982). Note seismic reflectors McDuff, Angus, Jimmy and "L". A seismic reflector locally corresponds to the Cretaceous-Tertiary hiatus on the Tugela Cone but has not been regionally correlated. Time scale is from Harland _et al._ (1982). Full names of Eras, Epochs and Ages: Valanginian, Hauterivian, Barremian, Aptian, Albian, Cenomanian, Turonian, Coniacian, Santonian, Campanian, Maestrichtian, Palaeocene, Eocene, Oligocene, Miocene, Pliocene, Quaternary, Pleistocene, Holocene.

The Upper Cretaceous outcrops in both northern and southern Natal as the St Lucia and Umzamba Formations (Figure 2.5) and is intersected in boreholes at Richards Bay, Durban and JC-1 (Du Toit and Leith, 1974; Kennedy and Klinger, 1975; Maud and Orr, 1975).

To the north, the St Lucia Formation lies unconformably on Lower Cretaceous rocks whereas in other localities it lies directly on pre-break-up rocks. As in the case of the Lower Cretaceous, the lower part of the sequence occurs only in Zululand. In Mozambique the basal part of the sequence ranges from continental to euxinic marine in character, whereas in Natal it is shallow marine.

A short hiatus separates the Cretaceous from the Tertiary both in the JC-1 borehole and at Richards Bay (Maud and Orr, 1975; McLachlan and McMillan, 1979). Although an acoustic reflector does coincide with this unconformity near JC-1 (Du Toit and Leith, 1974) it has been correlated with other seismic profiles only in the Tugela Cone area and has not been correlated regionally in the Natal Valley (Martin et al. 1982). Palaeocene and Eocene rocks in the JC-1 borehole comprise clays and sandy limestones and extend to the Eocene Oligocene boundary, whereas at Richards Bay only the basal Palaeocene is present. Elsewhere on the coastal plain of Natal, Lower Cretaceous rocks are absent whereas in southern Mozambique a variety of rocks include carbonates, clays and sands.

Near the Eocene/Oligocene border another hiatus occurs in JC-1, and is seen in south-western Mozambique but not in south-eastern Mozambique (Frankel, 1972; Flores, 1973; Forster, 1975). The JC-1 hiatus is associated with prominent regional acoustic reflector Angus (Dingle et al. 1978; Martin et al. 1982) which correlates with a long hiatus on the Mozambique Ridge separating Maastrichtian and Miocene open-ocean carbonates (Simpson, Schlich et al. 1974).

Late Tertiary rocks are not well developed on the Natal coastal plain. A shelly limestone in Zululand, the Pecten-bed of Late Miocene-Pliocene age (Siesser and Miles, 1979), is overlain by aeolianites which are in turn overlain by unconsolidated red sands and aeolianites of Pleistocene age (McCarthy, 1967; Maud, 1968; McLachlan and McMillan, 1979). These are equivalent to aeolianite cordons which form prominent shoals on the continental shelf such as Aliwal Shoals and Protea Bank (Figures 2.13 and 2.14). Both the Pleistocene Port

Durnford Formation and modern sediments of the Zululand coastal plain comprise beach barriers and aeolianites backed by lagoonal clays (Hobday and Orme, 1974; Hobday, 1979). In the JC-1 borehole, Oligocene rocks comprise sandy and shelly limestones but Miocene-Recent rocks were not continuously sampled (Du Toit and Leith, 1974). Offshore Late Miocene/Early Pliocene foraminiferal muddy sands underlie regional acoustic reflector Jimmy (Martin et al. 1982), while in DSDP hole 249 (Simpson et al 1974) much of the Pliocene is missing and the Quaternary is a condensed sequence.

DEEP STRUCTURE

Observation of the deep structure below the continental shelf requires powerful multi-channel seismic systems used predominantly by commercial companies. To date only one seismic profile has been released by the Southern Oil Exploration Corporation (Du Toit and Leith, 1974) (Figure 2.6). Elsewhere the deeper structure of the continental margin has been reconstructed from available seismic refraction and single-channel seismic reflection data (Dingle et al. 1978; Martin, 1984) (Figures 2.7 and 2.8). Although single-channel seismic systems are useful on the continental slope and the adjacent basin they do not reveal deep structure directly under the shelf because of multiple reflections and lack of penetration.

On the inner shelf in water-depths between 40 and 90 m, a series of seismic refraction stations parallel to the shore showed that deep sedimentary basins off Durban and Port St Johns are separated by a basement high at Port Shepstone (30°21'S-30°51'S), where basement is only 200 m below the seafloor (Ludwig et al. 1968). The sedimentary basin off Durban extends northward to the Mozambique coastal plain. North of 29°S, basement comprises Lebombo volcanic rocks which dip eastwards at angles of up to 50° before flattening out under the Natal Valley (Frankel, 1972; Beck and Lehner, 1974; Dingle et al. 1978; Martin, 1984) (Figure 2.7). Under the Tugela Cone basement is formed by Palaezoic rocks of the Natal Group which were encountered at the base of the JC-1 borehole. Basement dips gently eastwards, abutting oceanic crust of the deeper Natal Valley near the eastern face of the

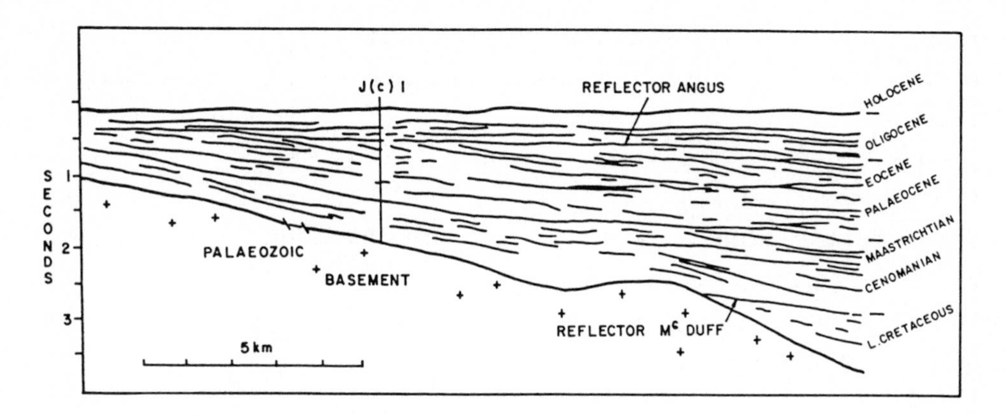

Figure 2.6 An interpretation of a seismic profile presented by Du Toit and Leith (1974), across the inner Tugela Cone. The JC-1 bore-hole provides stratigraphic control, with the location given on Figure 2.4.

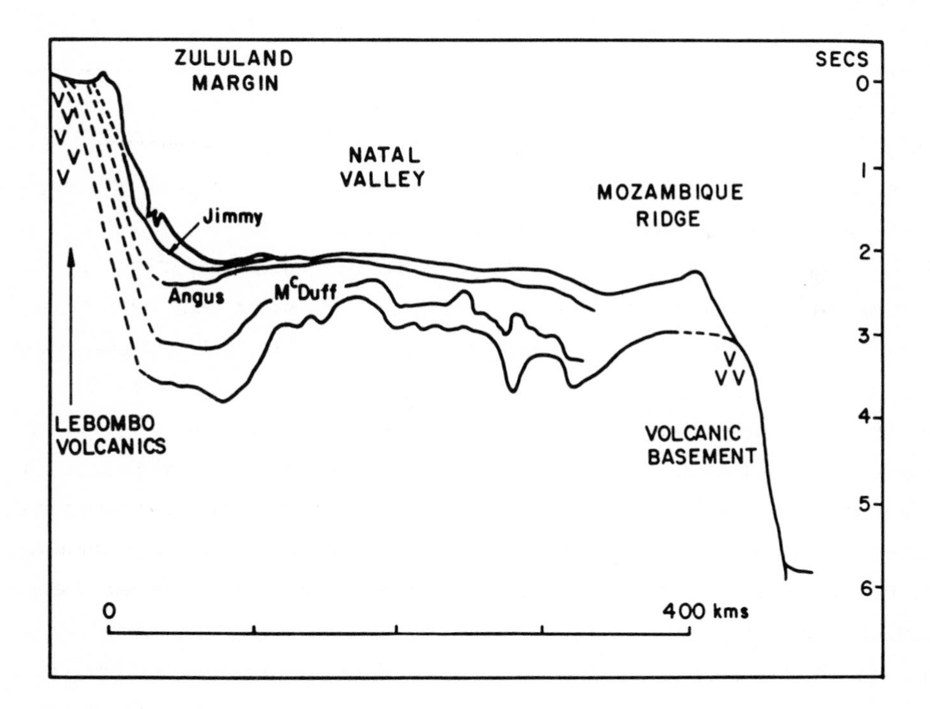

Figure 2.7 Geological section across the Zululand continental margin to the Mozambique Ridge, based on seismic profiles (Dingle <u>et al</u>. 1978; Martin, 1984) and coastal bore-holes (McLachlan and McMillan, 1979). Location on Figure 2.1.

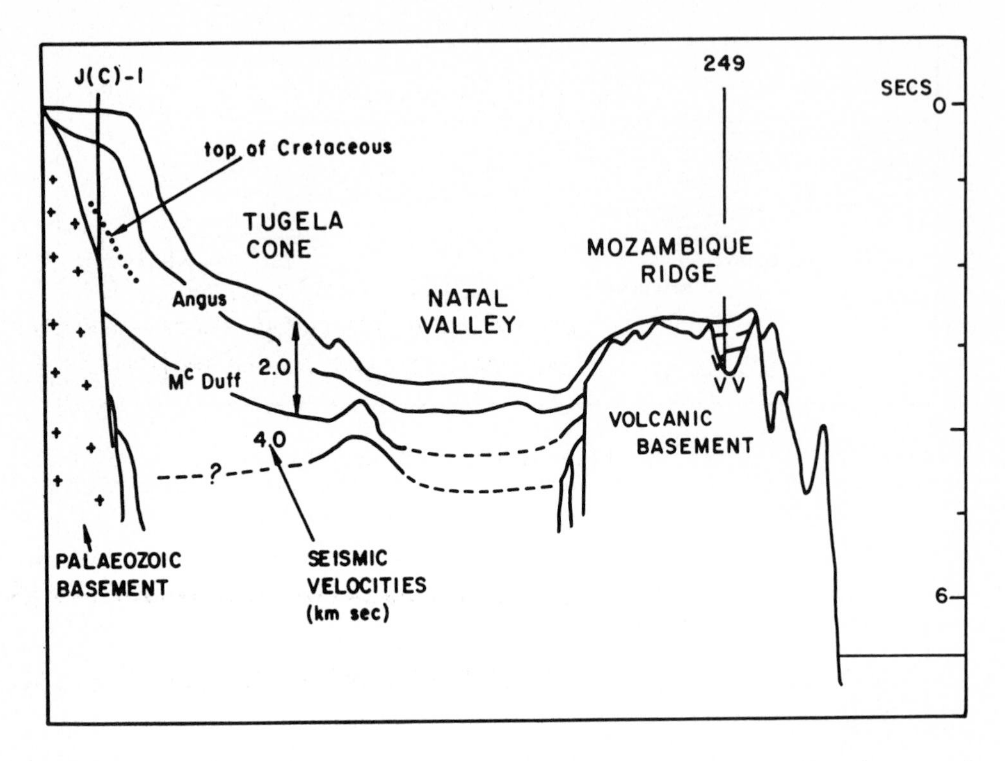

Figure 2.8 Geological cross-section across the Tugela Cone and the Mozambique Ridge (after Dingle <u>et al.</u> 1978).

Tugela Cone (Figures 2.6 and 2.8) (Du Toit and Leith, 1974; Chetty and Green, 1977; Dingle <u>et al</u>. 1978). Basement was extensively faulted during and after continental break-up in the Jurassic and Lower Cretaceous (Maud, 1961; Dingle <u>et al</u>. 1983, Figure 14.3).

There are two major depocentres in the area - the Limpopo and Tugela Cones (Figure 2.1). Maximum sediment thickness reaches 3.5 km along the Zululand margin and Limpopo Cone (Table 2.2 - note that values corrected for compaction are also given) (Beck and Lehner, 1974; Martin, 1984). The data of du Toit and Leith (1974) prove a sediment thickness of 4.7 km on the Tugela Cone, but it is suspected that 5.95 km exist (Figure 2.8). These depocentres are much shallower than other basins around Africa where up to 7, 8 and 12 km of sediment has accumulated in the Walvis, Orange and Outeniqua Basins (Dingle, 1982; Dingle <u>et al</u>. 1983).

Pre-Cenomanian acoustic reflector McDuff (Figure 2.5) caps a very reflective Early Cretaceous sedimentary unit which has been extensively faulted, with fault throws up to 945 m (Martin, 1984). Under the Tugela Cone, McDuff abuts basement east of the JC-1 borehole and pre-McDuff sediments thicken offshore to a known maximum of 1 700 m although a thickness of 2 600 m is suspected (Figures 2.6 and 2.8). Abundant volcanic ash on the Mozambique Ridge (Simpson, Shlich et al. 1974) and ubiquitous faulting evince an active tectonic volcanic régime.

The McDuff-Angus sequence (Figures 2.5 to 2.8) includes a Late Cretaceous sequence and an Early Tertiary sequence, and only under the Tugela Cone have these two been distinguished (Figures 2.5 to 2.7). In this area, both sequences dip eastwards and thicken to known maxima of 1 300 m and 1 580 m respectively (Table 2.2). In the JC-1 borehole, the Late Cretaceous is interpreted as an onlapping transgressive marine succession, whereas the Early Tertiary is a prograding subaqueous deltaic sequence (Du Toit and Leith, 1974). Under the Zululand margin, McDuff-Angus strata reach a maximum thickness of 1 823 m (Martin, 1984). This sequence of sediments infills the faulted irregular topography of the McDuff reflecting horizon, and therefore great variations in thickness occur. Many of the faults have been re-activated (growth faults), indicating variable loading and subsidence during sedimentation.

SHALLOW STRUCTURE

Because of the acquisition of high resolution seismic profiles and sidescan sonographs of over 5 000 km of seabed, the structure of the shallow continental shelf is better known than the deep structure (Flemming, 1981; Martin, 1985; Martin and Flemming, 1986). Several important characteristic features control the geology of the area (Figure 2.4).

Truncated seaward-dipping strata

Profiles running perpendicular to the shore generally exhibit sequences of strata dipping gently seaward. Several unconformities are evident, including an erosional unconformity apparently marking the

Table 2.2 Sediment thicknesses and accumulation rates. Sediment units relate to the stratigraphy given in Figure 2.5. CT = Cretaceous-Tertiary Boundary. Durations are from the time-scale of Harland *et al*. (1982). Note that 1) thickness values are followed by an accumulation rate in brackets; 2) maximum thicknesses of sediment units occur in different areas of the margin and therefore cannot be totalled; 3) corrections for compaction are made using a curve of porosity versus depth given by Steckler and Watts (1978). Data sources: Tugela Cone, Du Toit and Leith (1974); Goodlad (1979); more detail is available in Goodlad (1986); Zululand margin: Martin (1984); Average East African margin refers to average of Zambezi, Tanzania, Kenya and Somali Basins given by Dingle (1982) except that the corrected values have been recalculated.

	Sediment Unit	Duration (m.y.)	Thickness at Position where Total Thickness is maximum (m)	Accumulation Rate	Corrected for Compaction	Maximum Thickness of Sediment Unit	Corrected for Compaction
Tugela Cone	Post-Angus	38	650	(17)	727 (19)	1 430 (38)	1 751 (46)
	CT-Angus	27	1 490	(55)	2 171 (80)	1 580 (59)	2 407 (89)
	McDuff-CT	33	1 210	(37)	2 197 (67)	1 300 (39)	2 445 (74)
	Pre-McDuff	44	2 600	(59)	5 276 (120)	2 600 (59)	5 276 (120)
	Total	142	5 950	(42)av.	10 371 (73)av.		
Zululand Margin	Post-Jimmy	5	510	(102)	558 (112)	680 (136)	764 (153)
	Angus-Jimmy	33	890	(27)	1 140 (35)	890 (27)	1 140 (35)
	McDuff-Angus	60	900	(15)	1 434 (24)	1 823 (30)	2 499 (42)
	Pre-McDuff	44	1 100	(25)	2 020 (46)	2 200 (50)	3 524 (80)
	Total	142	3 400	(24)	5 152 (36)		
Average East African Margin	22.5- 5Ma	17.5	1 950	(111)	2 354 (135)		
	65 - 38Ma	27	1 650	(61)	3 023 (112)		
	98 - 65Ma	33	1 000	(30)	2 076 (63)		
	119 - 98	21	620	(30)	1 248 (59)		
	188 - 119	69	3 026	(44)	6 135 (89)		
	Total	183	8 246	(45)	14 836 (81)		

Miocene/Pliocene boundary (Figure 2.9). With the exception of the youngest part of the sequence at the shelf break (Figure 2.12), the strata have been truncated on their upper surface by sea-level changes and the resultant transgressions and regressions of seas across the shelf and coastal plain.

Shelf-edge slumps

Almost the entire shelf-edge and continental slope has been affected by seaward-slumping of large sediment masses (Figure 2.4). Slump-related features have been recognized both in high-resolution seismic profiles (Figure 2.10) and in air-gun profiles of the continental slope and the Natal Valley (Martin, 1984; Goodlad, 1986). In the upper reaches of the slump, largely intact blocks slide and rotate down-slope, whereas in the toe of the slump-mass, folds, faults, thrusts and jumbled chaotic bedding are characteristic. Between the slump-mass and stable material up-slope, in a valley or "tensional depression", older strata are exposed in a "glide-plane scar" (Lewis, 1971; Coleman and Garrison, 1977; Dingle 1980). This occurs off Natal, where Cretaceous and Palaeogene rocks are exposed in glide-plane scars and canyon walls (Figure 2.4).

Slump features are exposed both on the Zululand margin (the Zululand slump described by Martin, 1984) as well as south of Durban, but on the broader shelf east of Durban the slump sequence is overlain by younger material. It is uncertain whether the Zululand slump comprises several separate slumps or one large mass, but slump-related features extend over an area of at least 20 800 km^2, with the slump-mass being up to 520 ms (416 m) thick. In size the Zululand slump rivals other enormous slumps on the southern African margin which are the largest such features yet described (Dingle, 1980).

Canyons

Several canyons dissect the continental slope off Natal (Figure 2.2) and these have acted as conduits for sediment transport from the shelf to the deep sea (Martin, 1984; Goodlad, 1986). On the Zululand slope, canyons cut across the Zululand slump (Figure 2.11) and in some cases

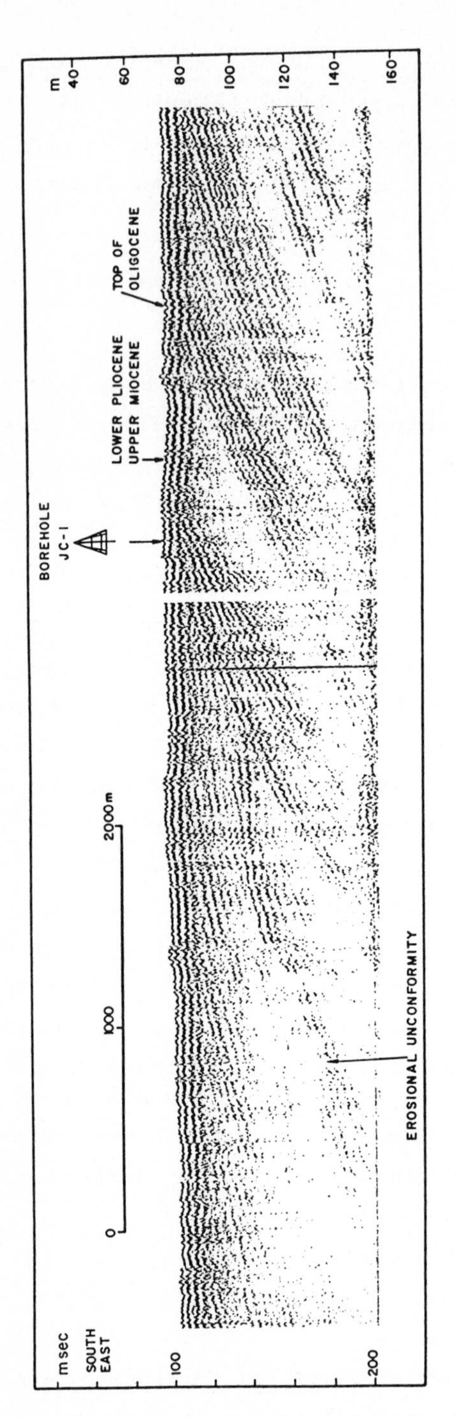

Figure 2.9 A sparker seismic profile showing truncated seaward-dipping strata. Note an erosional unconformity which correlates to the Miocene-Pliocene boundary hiatus which is called Jimmy in the Natal Valley (Figure 2.5). The location is given on Figure 2.4.

there may be a relationship between canyons and slippage of large sediment masses. On the Tugela Cone, the inshore ends of large canyons do not reach the shelf edge, but both to the north and south on narrow

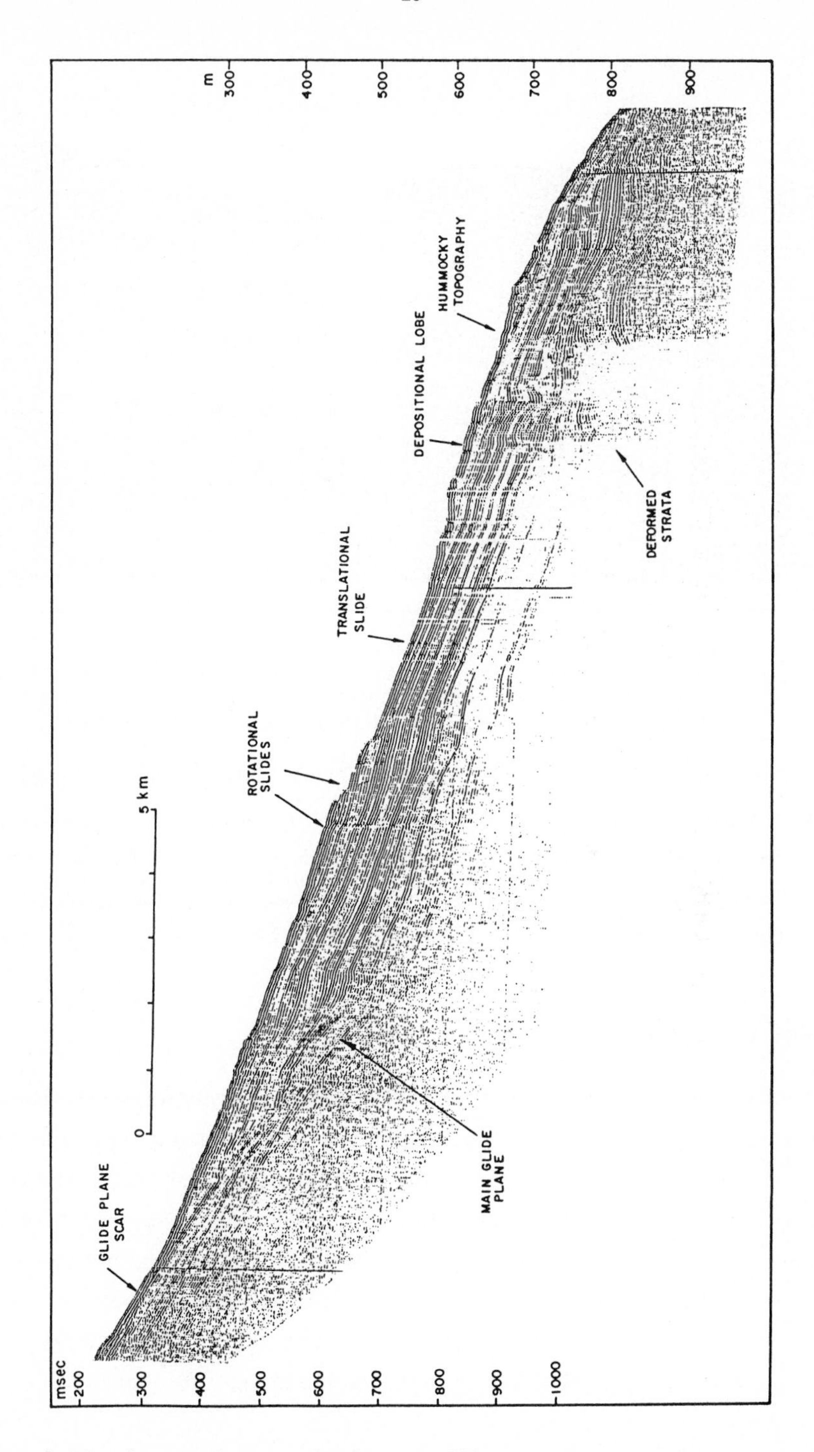

Figure 2.10 A sparker seismic profile across a slump south of Durban. The location is given on Figure 2.4.

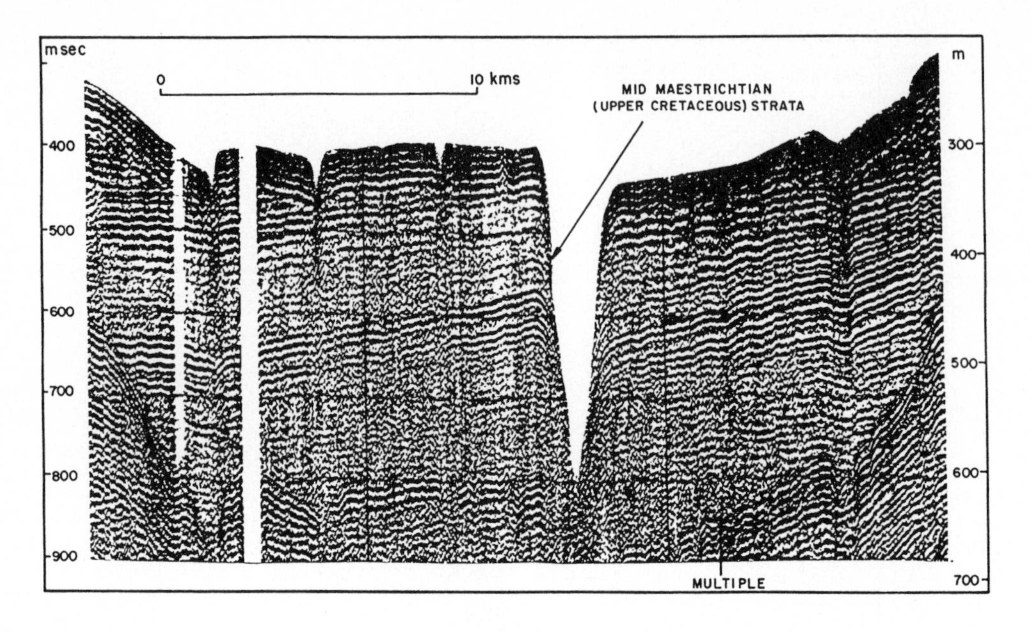

Figure 2.11 A single-channel air-gun seismic profile across Leven Canyon, where Late Cretaceous strata outcrop on the canyon wall. The location is given on Figure 2.4.

sections of shelf, canyons penetrate the shelf edge (Figure 2.4) (Flemming, 1981; Martin and Flemming, 1986). In such cases, unconsolidated Holocene sediments which are entrained by the powerful Agulhas Current are funnelled into canyon-heads and transported to the deep-sea by turbidity currents (Chapter 3).

Shelf-edge prograding sequence

Along parts of the Zululand shelf, and particularly on the broad shelf east of Durban, prograding sequences form the shelf-break (Figure 2.12). Unlike their older counterparts further inshore, these units are not truncated at their upper surfaces. Because they overlie slumps which have deformed Pliocene strata, these prograding bodies must be Late Pliocene or Pleistocene. Their internal structure displays complex sigmoid oblique clinoforms (Vail *et al*. 1977) suggesting a complex pattern of outbuilding and upbuilding of sediment (Figure 2.12).

The observed geological distribution (Figure 2.4) results from the

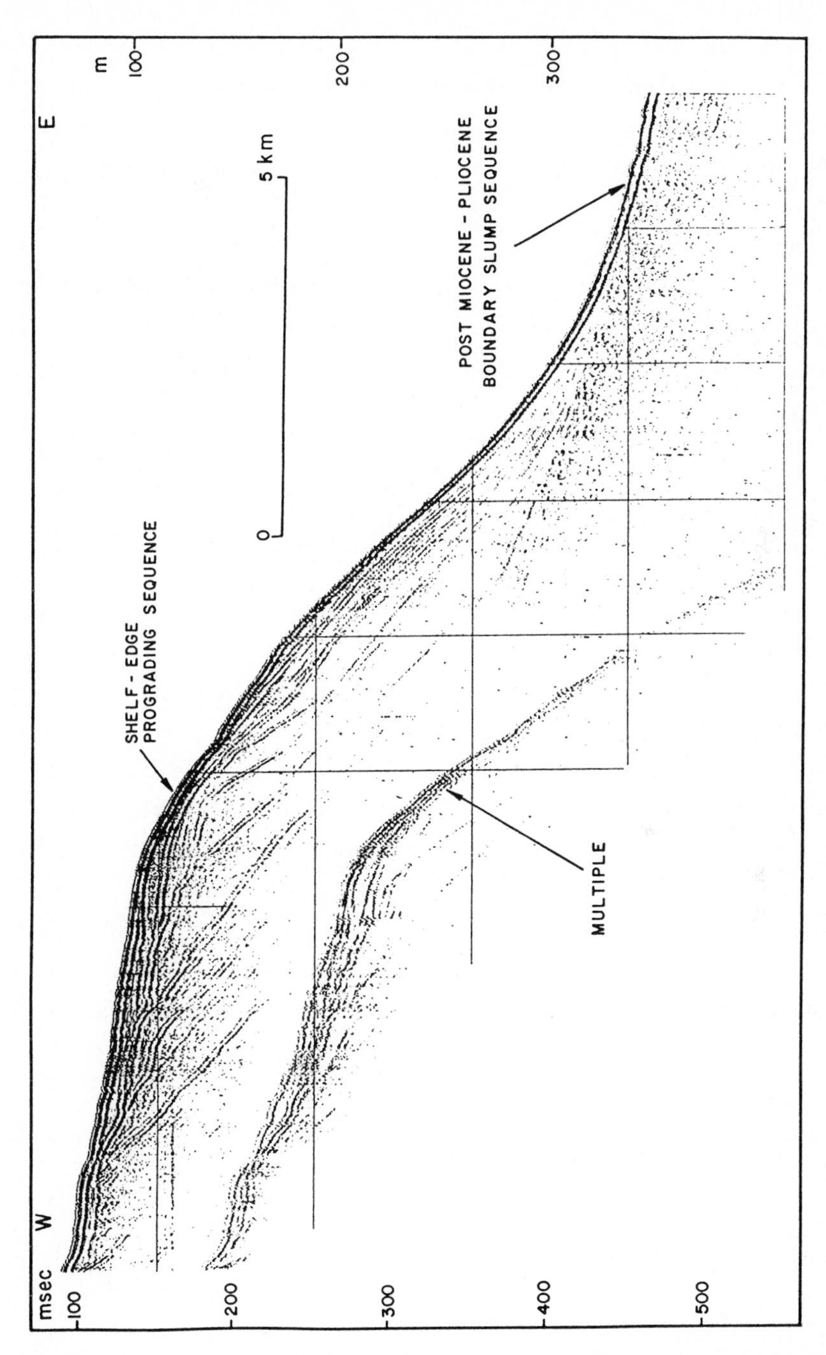

Figure 2.12 A sparker seismic profile showing a seaward-prograding sequence at the shelf-edge east of Durban. Note that this sequence overlies contorted strata of a slumped sequence. The location is given on Figure 2.4.

interplay of the above features. Older truncated sequences outcrop inshore, while successively younger units outcrop to seaward. The exceptions are Cretaceous and Palaeocene strata exposed in glide-plane scars and canyon walls.

Pleistocene Formations

Three Pleistocene-Recent sequences unconformably overlie truncated seaward-dipping strata: a) aeolian foredune complexes; b) fluvial sediments infilling buried channels; c) unconsolidated Holocene sediments (Chapter 3).

a) Aeolianites. Rugged, linear, aeolianite shoals form some of the most prominent features of the Natal shelf (Figure 2.13). In places, as at Protea Banks and Aliwal Shoal, these aeolianites are serious hazards to shipping. Onshore and offshore aeolianites extend from Mozambique to the western Cape (McCarthy, 1967; Maud, 1968; Davies, 1976; Birch, 1981; Flemming, 1981; Barwis and Tankard, 1983; Martin and Flemming, 1986). On the south coast of South Africa, observation by scuba-divers and on-land field-work shows that the aeolianites are associated with beach, wash-over fan, lagoonal and estuarine facies, pointing to their formation as coastal dunes associated with barrier beaches (Martin and Flemming, 1986). Similarly in Natal, by analogy with on-land occurrences, and from the examination of aeolianite samples taken from Aliwal Shoal during a shipping accident (McCarthy, 1967), these linear shoals (Figure 2.14) are interpreted as submerged coastal dune cordons. Re-evaluation of micropalaeontological data (McLachlan and McMillan, 1979) and available C^{14} dates (Deacon, 1966; Barwis and Tankard, 1983) suggest formation of the shoals during the last interglacial between 120 000 and 30 000 years ago.

b) Fluvial channel infill. Incised valleys on the present continental shelf are observed in seismic profiles off many Natal rivers (Figure 2.15). These valleys were cut during sea-level low-stands when river courses extended to the continental slope. The base of the buried channel off the Tugela River descends from -87 m 10 km offshore to -104 m 34 km offshore. It is likely that thisdowncutting occurred during glacial periods such as occurred 18 000 years ago when sea-level

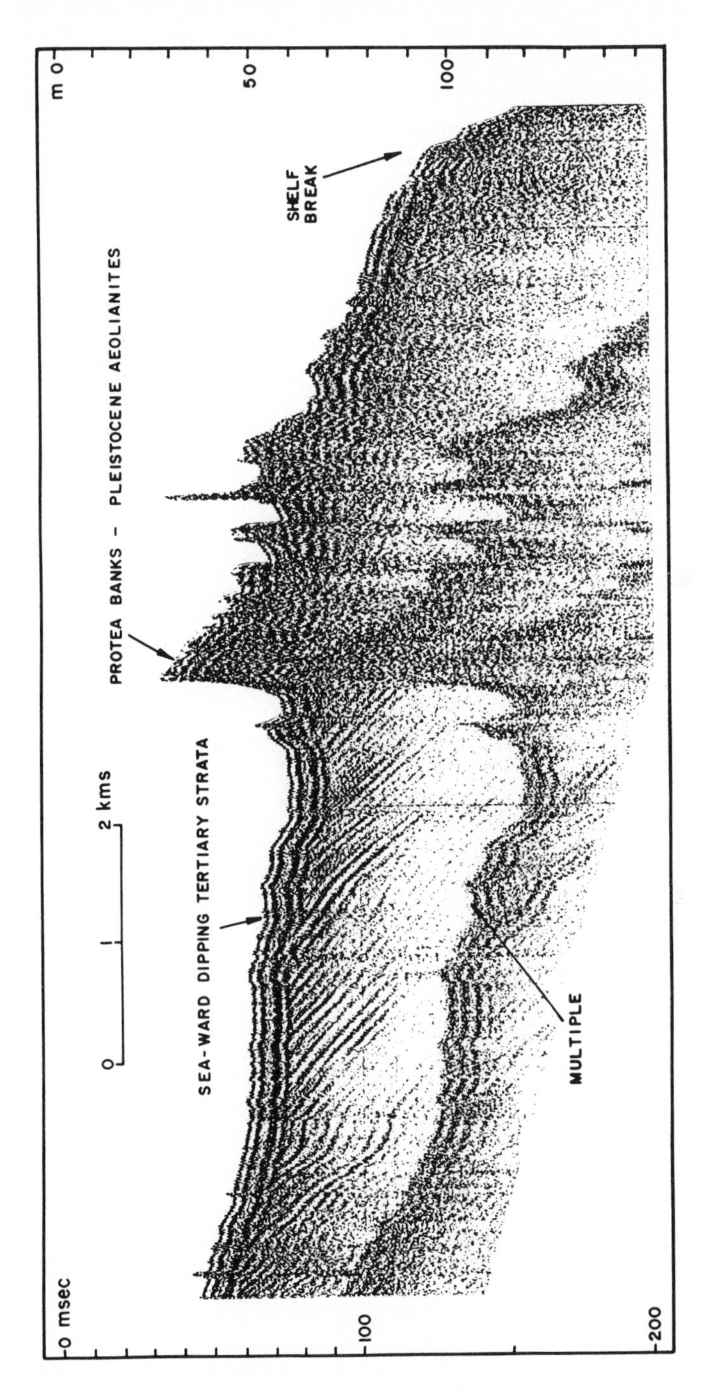

Figure 2.13 A sparker seismic profile across Protea Banks, part of a Pleistocene aeolianite cordon. Note that the aeolianite ridge unconformably overlies truncated seaward-dipping strata. The location is given on Figure 2.14.

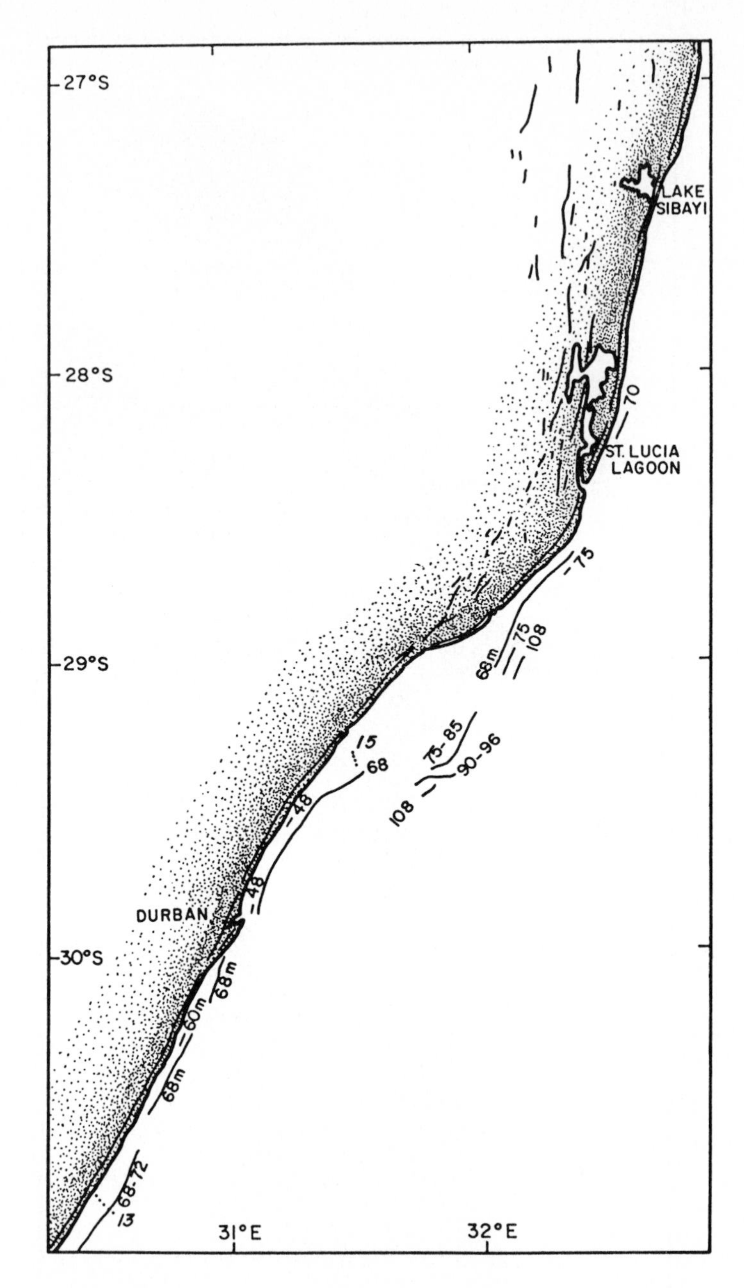

Figure 2.14 Pleistocene aeolianite cordons of the Natal continental shelf mapped from over 5 000 km of seismic profiles. The numbers refer to the depth of the base of the cordon which marks the inferred sea-level. Onshore aeolianites of the coastal plain are associated with high sea-levels (after Davies, 1976). Dotted lines 13 and 15 are locations of Figures 2.13 and 2.14).

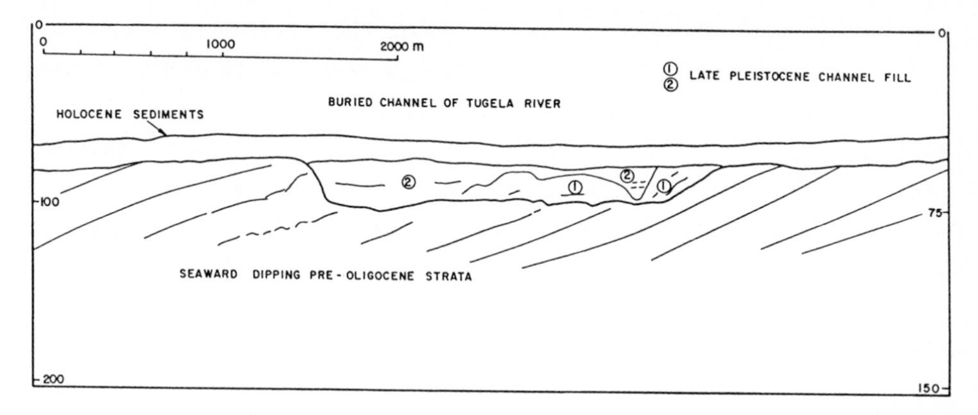

Figure 2.15 An interpretation of a sparker seismic profile across the buried channel of the Tugela River. Note the successive fill of the channel marked as units 1 and 2, overlain by unconsolidated Holocene sediments. The location is given on Figure 2.14.

was probably 130 m below present sea-level (Figure 2.17). Channels are filled by successive sedimentary units resembling point bars and are overlain by Holocene sediments deposited when sea-level regained levels near to those of present day.

FACTORS INFLUENCING CONTINENTAL MARGIN EVOLUTION

Terrestrial sediment supply

Studies of deep-sea drilling sites worldwide show that phases of fast and slow sedimentation have alternated in Cretaceous-Recent times (Davies et al. 1977). Subsequent work (Worsley and Davies, 1979), which took sediment compaction into account, suggested that rates vary by a factor of 5, and that high rates in the ocean basins are related to low sea-levels.

In Natal, modern rivers supply large quantities of sediment to the sea (Flemming, 1981, and Chapter 3). The present-day average denudational rate for the east coast drainage basin is 215 tonnes km^{-2}, an unexceptional rate for Africa (Walling, 1984). Estimates of sediment accumulation rates in the entire Natal Valley averaged over the last 5 Ma and over the last 130 Ma indicate that modern rates of supply are between 12 and 22 times higher than rates averaged over geologic time (Martin, 1986). This suggests that in the geologic past, average erosion rates were of the order of 10-18 tonnes km^{-2}. In the eastern United States agricultural malpractices may have caused a ten-fold increase in erosion in recent times (Meade, 1982). By comparison, the scale of increase in southern Africa is cause for concern.

Subsidence and sediment loading

Sediment accumulation at any one spot has been shown to be predominantly controlled by subsidence (Watts, 1982). Tectonic subsidence results from cooling of the lithosphere after heating, stretching and thinning at the time of rifting (McKenzie, 1978; Steckler and Watts, 1978). In addition, subsidence is driven by the load of accumulated sediment and overlying water. Initially, subsidence occurs more locally through block faulting, and therefore sedimentary basins are narrow and deep (Airy-type isostasy). Later in continental-margin evolution, as the lithosphere cools and gains mechanical strength, subsidence occurs over a broader area, so that basins become wider and deepen at a lower rate (flexural isostasy) (Watts, 1982).

Accumulation rates in the Tugela Cone and the Zululand continental margin show contrasting patterns (Table 2.2). On the Tugela Cone, sedimentation rates are high initially, and decrease with time. This is in accord with its position adjacent to a spreading ridge segment where a high heat anomaly is expected at the time of rifting. On the Zululand continental margin, and in east African basins as a whole, initial rates are moderately high, decrease in the Late Cretaceous, and increase again in Late Tertiary-Recent times. Subsidence history from the Early Cretaceous to the Early Tertiary is commensurate with the decay of a rift-related heating anomaly, but renewed subsidence in the Late Tertiary-Recent is not. Deep-sea sedimentation rates both locally and globally mirror the increase in sedimentation rates in the Late Tertiary (Girdley _et al_ 1974; Davies _et al_. 1977; Worsley and Davies, 1979). Various reasons for this have been given, but increasing contents of aluminium oxide suggest increased terrestrial weathering associated with cooler climate (Donnelly, 1982). This does not, however, explain the increase in subsidence, although possible tectonism in the last 10 Ma may provide an answer (see Hinterland Tectonics).

Sea-level (Eustasy)

Global sea-level is controlled by the volume of the ocean basins and the quantity of water locked up in terrestrial ice-caps. Calculations of ocean basin volumes, which depend primarily on the spreading rates and volumes of mid-ocean ridges (Vail et al. 1977; Pitman, 1978), suggest that absolute sea-level should have fallen steadily since the mid-Cretaceous. Relative sea-level depends on the rate at which sea-level drops compared with the rate of subsidence and sedimentation on continental margins. Since the Oligocene, Antarctic ice volumes (glacio-eustasy) as well as tectono-eustasy have affected sea-level. Tectono-eustatic changes occur at rates less than 1 cm/1 000 yr, glacio-eustatic changes occur at up to 1 000 cm/1 000 yr, whereas margin subsidence proceeds at 2 cm/1 000 yr (Pitman, 1978).

A sea-level curve for southern Africa (Figure 2.16) has been constructed from the altitudes at which strata of various ages abut or onlap older sequences (Siesser and Dingle, 1981; Dingle et al. 1983). This curve, therefore, matches the stratigraphic column, with hiatuses in onland and shelf sequences correlating with low sea-levels. Given the above arguments, the curve displays relative rather than absolute sea-level. For example, on the Tugela Cone, Early Cretaceous rocks (pre-McDuff) abut basement at -2 800 m (Figure 2.8), whereas rocks of this age are exposed on the Zululand coastal plain (Figure 2.4).

A more detailed curve of global sea-levels (Figure 2.17) has been constructed for the Late Pleistocene during which time dramatic sea-level changes have resulted from climatic oscillations and changing ice volumes (Williams et al. 1981). In Natal, the aeolianite cordons mapped in Figure 2.14 represent fossil strandlines, and notches near the shelf-edge at -115 m may represent sea-cliffs eroded at low sea-level. A program of sample dating is required to compare these sea-level indicators with existing sea-level curves. Similarly, the

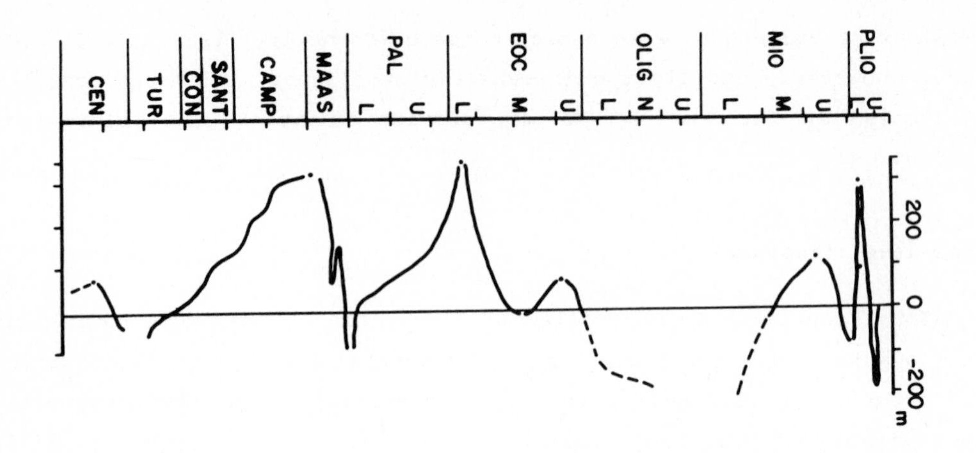

Figure 2.16 A relative sea-level curve for Southern Africa during the Late Cretaceous to Quaternary (after Siesser and Dingle, 1981; Dingle *et al.* 1983).

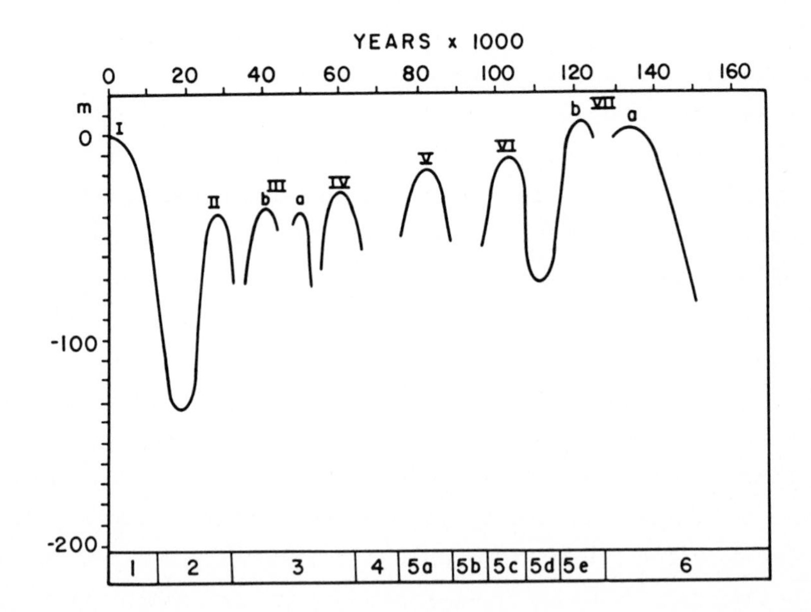

Figure 2.17 Sea-level curve for the last 150 000 years relative to present sea-level. Note high sea-levels I to VII a and oxygen isotope stages 1-6 (after Williams *et al.* 1981).

various positions of coastal onlap displayed by prograding shelf-edge sediment bodies (Figure 2.12) record relative sea-levels between -100 and -200 m.

Hinterland tectonics

Repeated uplift of the Natal hinterland rather than eustatic sea-level changes has been invoked to explain the periodic emergence of the continental shelf (King, 19172). Although this argument has been criticised (Frankel, 1972; De Swardt and Bennett, 1974), the 3 000 m altitude of flat-lying Karoo rocks in the Drakensburg Mountains does require explanation. Karoo volcanism including that of the Lebombo Mountains, the possible involvement of hot-spots, and the break-up of the continents doubtless caused uplift in the Jurassic (Martin and Hartnady, 1986). In the last 10 Ma, uplift of "high" Africa may have occurred by two mechanisms. Smith (1982) suggested that the African continent moved over a heat anomaly in the asthenosphere (at 150 km depth) which resulted in a phase-change at the base of the crust and produced concomitant uplift. Hartnady (1985) has suggested that a hot-spot presently lies under the Drakensberg Mountains and that epeirogenic uplift resulted as first the Natal Valley then coastal Natal drifted over the hot-spot. The gradual decay of such a heating anomaly may be the cause of renewed subsidence of the Natal continental margin in the Late Tertiary-Recent (Table 2.2).

Palaeo-Oceanography

The Agulhas Current has influenced sedimentation both on the continental shelf (Chapter 3) and in the Natal Valley (Dingle et al. 1978; Goodlad, 1979; Martin, 1981a, 1981b, 1984; Martin et al. 1982). Initial sediments deposited in the Indian and Southern Oceans were euxinic, and the change to open-marine conditions was likely related to the on-set of oxygenated bottom-water flow (Schlanger and Jenkyns, 1976; Andrews, 1977; Natland, 1978; Brass et al., 1982). In the Natal Valley, sedimentary bodies immediately underlying reflector Angus show clear evidence for current-moulding, and confirm the influence of the Agulhas Current. Invigoration of sub-tropical oceanic gyres is linked to the initiation of cold dense bottom currents in the world ocean near the Eocene/Oligocene boundary (Kennett and Shackleton, 1976). New Agulhas Current flow-paths similar to those of today were established in post-Jimmy times (5 Ma - see Figure 2.5). This is associated with new global patterns of bottom-water production related to the sinking of cold dense water around both the Antarctic and Arctic ice-caps (Shackleton and Opdyke, 1977; Moore et al. 1978; Kaneps, 1979; Herman and Hopkins, 1980). Pleistocene fluctuations in the Agulhas Current are linked to glacial/interglacial oscillations (Hutson, 1980).

Such large-scale changes in the Agulhas Current regime must have directly altered the ecology of the Natal coastal ocean by influencing sea temperature and circulation patterns and indirectly by affecting climate. Changes in current patterns in Oligocene and Pliocene times likely resulted in more inflow of cooler sub-tropical water to the ocean off south-eastern Africa (Martin, 1981b). This would have reduced sea-surface evaporation, leading to less moisture in air inflowing over sub-tropical southern African and reduced rainfall in these areas. These climatic effects were contributary causes of faunal turnovers in southern Africa occurring in the Early Oligocene and at the Miocene/Pliocene boundary.

Acknowledgements

Unpublished data appearing here was collected as part of the NRIO's sediment/current interaction project, aided by Jack Engelbrecht, Gerry Rule and Walter Akkers. Thanks are also due to Richard Dingle for making available details of dredged samples.

REFERENCES

ANDERSON, W (1907). Third and final report of the Geological Survey of Natal and Zululand. Surveyor General Department Natal Colony.

ANDREWS, P B (1977). Depositional facies and the early phase of ocean basin evolution in the circum-Antarctic region. **Marine Geology,** 25 1/3, 1-13.

BARRON, E J, C G A HARRISON and W W HAY (1978). A revised re-construction of the southern continents. **EOS Transactions of the American Geophysical Union,** 59, 436-449.

BARWIS, J H and A J TANKARD (1983). Pleistocene shoreline deposition and sea-level history at Swartklip, South Africa. **Journal of Sedimentary Petrology**, 53, 1281-1294.

BECK, R H and P LEHNER (1974). Oceans, new frontiers in exploration. **American Association of Petroleum Geologists Bulletin,** 53, 376-395.

BIRCH, G F (1981). The bathymetry and geomorphology of the continental shelf and upper slope between Durban and Port St Johns. **Annals of the Geological Survey of South Africa,** 15/1, 55-62.

BRASS, G W, J R SOUTHAM and W H PETERSON (1982). Warm saline bottom water in the ancient ocean. **Nature,** 296, 620-623.

CHETTY, P and R W E GREEN (1977). Seismic refraction observations in the Transkei Basin and adjacent areas. **Marine Geophysical Researches,** 3, 197-208.

COLEMAN, J N and L E GARRISON (1977). Geological aspects of marine slope instability, northwestern Gulf of Mexico. **Marine Geotechnology,** 9-44.

DAVIES, O (1976). The older coastal dunes in Natal and Zululand and their relation to former shorelines. **Annals of the South African Museum,** 71, 19-32.

DAVIES, T A, W W HAY, J R SOUTHAM and T R WORSLEY (1977). Estimates of Cenozoic oceanic sedimentation rates. **Science,** 197, 53-55.

DEACON, H J (1966). The dating of the Nahoon footprints. **South African Journal of Science,** 62, 111-113.

DE SWARDT, A M J and G BENNETT (1974). Structural and physiographic development of Natal since the Late Jurassic. **Transactions of the Geological Society of South Africa,** 77, 309-322.

DINGLE, R V (1980). Large allochthonous sediment masses and their role in the construction of the continental slope and rise off southwestern Africa. **Marine Geology,** 37, 333-354.

DINGLE, R V (1982). Continental margin subsidence: a comparison between the east and west coasts of Africa. In: **Dynamics of passive margins,** (Ed.) R A SCRUTON, American Geophysical Union, Geodynamics Series 6, 59-71.

DINGLE, R V and W G SIESSER (1977). Geology of the continental margin between Walvis Bay and Ponto do Ouro. Geological Survey of South Africa, **Marine Geoscience Map series 2.**

DINGLE, R V, S W GOODLAD and A K MARTIN (1978). Bathymetry and stratigraphy of the northern Natal Valley (SW Indian Ocean). A preliminary report. **Marine Geology,** 28, 89-106.

DINGLE, R V, W G SIESSER and A R NEWTON (1983). **Mesozoic and Tertiary geology of southern Africa.** A.A. Balkema, Rotterdam, 375 pp.

DONNELLY, T W (1982). Worldwide continental denudation and climatic deterioration during the late Tertiary: evidence from deep-sea sediments. **Geology** 10, 451-454.

DU TOIT, A L (1937). **Our wandering continents.** Oliver and Boyd. Edinburgh, 366 p.

DU TOIT, S R and M J LEITH (1974). The J(C)-1 bore-hole on the continental shelf near Stanger, Natal. **Transactions of the Geological Society of South Africa,** 77, 247-252.

DUNCAN, R A (1984). Age progressive volcanism in the New England seamounts and the opening of the Central Atlantic Ocean. **Journal of Geophysical Research,** 89, 9980-9990.

FLEMMING, B W (1981). Factors controlling shelf sediment dispersal along the south-east African continental margin. **Marine Geology,** 42, 259-277.

FLORES, G (1973). The Cretaceous and Tertiary sedimentary basins of Mozambique and Zululand. In: Sedimentary Basins of the African Coasts. (Ed.) G BLANT. **Association f African Geological Surveys, Paris,** 81-111.

FORSTER, R (1975). Geological history of the sedimentary basin of southern Mozambique and some aspects of the origin of the Mozambique channel. **Palaeogeography, palaeoclimatology, and palaeo-ecology,** 17, 267-287.

FRANKEL, J J (1972). Distribution of Tertiary sediments in Zululand and Southern Mozambique, Southeast Africa. **American Association of Petroleum Geologists Bulletin,** 56, 2415-2425.

GIRDLEY, W A, L LECLAIRE, C MOORE, T L VALLIER and S M WHITE (1974). Lithologic Summary, Leg. 25, Deep-Sea drilling project. In: **Initial Report of the Deep-Sea Drilling Project 25,** E S W SIMPSON, R SCHLICH, et al., Washington (U.S. Government Printing Office), 725-741.

GOODLAD, S W (1979). Some aspects of deep current activity in the mid-Natal valley. **Joint Geological Survey/University of Cape Town Marine geology programme technical report,** 11, 91-99.

GOODLAD, S W (1986). A tectonic and sedimentary history of the mid-Natal Valley (SW Indian Ocean). **Joint Geological Survey/University of Cape Town Marine Geology programme bulletin,** 15, 415 pp.

HARLAND, W B, A V COX, P G LLEWELLYN, C A G PICKTON, A G SMITH and R WALTERS (1982). **A geologic time-scale.** Cambridge University Press, 131 pp.

HARTNADY, C J H (1985). Uplift, faulting, seismicity, thermal spring and possible incipient volcanic activity in the Lesotho-Natal region, SE Africa, the Quathlamba hotspot hypothesis. **Tectonics,** 4, 371-377.

HAY, E R (1984). Sediment dynamics on the continental shelf between Durban and Port St Johns (southeast African continental margin). **Joint Geological Survey/University of Cape Town Marine geology programme bulletin,** 13, 238 pp.

HAY, W W, E J BARRON, J L SLOAN and J R SOUTHAM (1981). Continental drift and the global pattern of sedimentation. **Geologisches Rundschau,** 70(1): 302-315.

HERMAN, Y and D M HOPKINS (1980). Arctic Oceanic climate in Late Cenozoic time. **Science,** 209, 557-563.

HOBDAY, D K (1979). Geological evolution and geomorphology of the Zululand coastal plain. In: **Lake Sibaya,** (Ed.) B R ALLANSON, Monographiae Biologicae 36, 1-20.

HOBDAY, D K (1982). The southeast African margin. In: **The Ocean basins and margins. 6: The Indian Ocean,** (Eds.) A E M NAIRN and F G STEHLI, pp 149-183.

HOBDAY, D K and A R ORME (1974). The Port Durnford Formation: a major Pleistocene barrier lagoon complex along the Zululand coast. **Transactions of the Geological Society of South Africa,** 77, 141-149.

KENNEDY, W J and H C KLINGER (1975). Cretaceous fauns from Zululand and Natal, South Africa. Introduction, Stratigraphy. **Bulletin Museum of Natural History. Geology,** 25, 265-315.

KENNETT, J P and N J SHACKLETON (1976). Oxygen isotopic evidence for the development of the psychrosphere 38 myr ago. **Nature,** 260, 513-515.

KING, L C (1972). **The Natal Monocline: explaining the origin and scenery of Natal, South Africa.** University of Natal, 112 pp.

LE PICHON, X and J C SIBUET (1981). Passive margins: a model of formation. **Journal of Geophysical Research,** 86, 3708-3720.

LEWIS, K B (1971). Slumping on a continental slope inclined at 1°-4°. **Sedimentology** 16, 97-110.

LUDWIG, W J, J E NAFE, E S W SIMPSON and S SACKS (1968). Seismic refraction measurements on the southeast African Continental Margin. **Journal of Geophysical Research,** 73, 3707-3719.

MARTIN, A K (1981a). The influence of the Agulhas Current on the physiographic development of the northernmost Natal Valley (SW Indian Ocean). **Marine Geology,** 39, 259-276.

MARTIN, A K (1981b). Evolution of the Agulhas Current and its palaeo-ecological implications. **South African Journal of Science,** 77, 547-554.

MARTIN, A K (1984). Plate tectonic status and sedimentary basin in-fill of the Natal Valley (SW Indian Ocean). **Joint Geological Survey/University of Cape Town Marine Geology Programme Bulletin** 14, 209 pp.

MARTIN, A K (1986). A comparison of sedimentation rates in the Natal

Valley, S.W. Indian Ocean, with modern sediment yields in east coast rivers, Southern Africa. **South African Journal of Science,** (in press).

MARTIN, A K and B W FLEMMING (1986). The Holocene shelf sediment wedge off the south and east coast of South Africa. **Canadian Society of Petroleum Geologists Memoir,** (in press).

MARTIN, A K and C J H HARTNADY (1986). Plate tectonic development of the south-west Indian Ocean: a revised reconstruction of East Antarctica and Africa. **Journal of Geophysical Research,** 91, 4767-4786.

MARTIN, A K, S W GOODLAD and D A SALMON (1982). Sedimentary basin in-fill in the northernmost Natal Valley, hiatus development and Agulhas Current palaeo-oceanography. **Journal of the Geological Society, London,** 139, 183-201.

MARTIN, A K, S W GOODLAD, C J H HARTNADY AND A DU PLESSIS (1982). Cretaceous palaeopositions of the Falkland Plateau relative to southern Africa using Mesozoic seafloor spreading anomalies. **Geophys. J. Roy. Astr. Soc.** 71, 567-579.

MAUD, R R (1961). A preliminary review of the structure of coastal Natal. **Transactions and Proceedings of the Geological Society of South Africa,** 64, 247-256.

MAUD, R R (1968). Quaternary Geomorphology and Soil Formation in Coastal Natal. **Zeitschrift für Geomorphologie supplementband,** 7, 153-199.

MAUD, R R and W N ORR (1975). Aspects of post-Karoo Geology in the Richards Bay area. **Transactions of the Geological Society of South Africa,** 78, 101-109.

McCARTHY, M J (1967). Stratigraphical and sedimentological evidence from the Durban Region of major sea-level movements since the late Tertiary. **Transactions of the Geological Society of South Africa,** 70, 135-165.

McKENZIE, D (1978). Some remarks on the development of sedimentary basins. **Earth and Planetary Science Letters,** 40, 25-32.

McLACHLAN, I R and I K McMILLAN (1979). Microfaunal biostratigraphy, chronostratigraphy and history of Mesozoic and Cenozoic deposits of the coastal margin of South africa. **Geological Society of South Africa Special publication,** 6, 161-181.

MEADE, R H (1982). Sources, sinks and storage of river sediment in the Atlantic drainage of the United States. **Journal of Geology,** 90, 235-252.

MOIR, G J (1974). Bathymetry of the Upper Continental margin between Cape Recife (34°S) and Ponta do Ouro (27°S) South Africa. **Joint Geological Survey/University of Cape Town Marine geology pro-**

gramme technical report, 7, 68-78.

MOORE, T C, T H VAN ANDEL, C SANCETTA and N PISIAS (1978). Cenozoic hiatuses in pelagic sediments. **Micropalaeontology,** 24(2), 113-138.

NATLAND, J H (1978). Composition, provenance and diagenesis of Cretaceous clastic sediment drilled on the Atlantic continental rise of Southern Africa, DSDP site 361 - Implications for the early circulation of the South Atlantic. In: **Initial Reports of the Deep-Sea Drilling Project 40,** H M BOLLI, W B F RYAN, et al., Washington, U.S. Government Printing Office, 1025-1061.

PITMAN, W C (1978). Relationship between eustacy and stratigraphic sequences of passive margins. **Bulletin of the Geological Society of America,** 89, 1389-1403.

SCHLANGER, S O nd H C JENKYNS (1976). Cretaceous oceanic anoxic events: Causes and consequences. **Geologie en Mijnbouw,** 55, 179-185.

SCRUTTON, R A (1979). On sheared passive continental margins. **Tectonophysics,** 59, 293-305.

SCRUTTON, R A (ED.) (1982). Dynamics of passive margins. **American Geophysical Union Geodynamics Series,** 6.

SHACKLETON, N J and N D OPDYKE (1977). Oxygen isotope and palaeomagnetic evidence for early northern hemisphere glaciation. **Nature,** 270, 216-219.

SHEPARD, F P (1973). **Submarine geology.** Harper and Row, New York, 517 pp.

SIESSER, W G (1977). Biostratigraphy and micropalaeontology of continental margin samples. **Joint Geological Survey/University of Cape Town Marine geology programme technical report,** 9, 108-117.

SIESSER, W G and R V DINGLE (1981). Tertiary sea-level movements around Southern Africa. **Journal of Geology,** 89, 83-96.

SIESSER, W G and G A MILES (1979). Calcareous nannofossils and planktic foraminifers in Tertiary Limestones, Natal and Eastern Cape, South Africa. **Annals of the South African Museum,** 79(6), 139-158.

SIMPSON, E S W, R SCHLICH, et al (1974). **Initial reports of the deep-sea drilling project 25.** Washington D.C., U.S. Government Printing Office, 884 pp.

SIMPSON, E S W, J G SCLATER, B PARSONS, I NORTON and L MEINKE (1979. Mesozoic magnetic lineations in the Mocambique Basin. **Earth and Planetary Scienc Letters,** 43, 260-264.

SMITH, A G (1982). Late Cenozoic uplift of stable continents in a reference frame fixed to South America. **Nature,** 296, 400-404.

STECKLER, M S and A B WATTS (1978). Subsidence of the Atlantic-type continental margin off New York. **Earth and Planetary Science Letters,** 41, 1-13.

TANKARD, A J, M P A JACKSON, K A ERIKSSON, D K HOBDAY, D R HUNTER and W E L WINTER (1982). **Crustal evolution of southern Africa.** Springer-Verlag, New York, 523 pp.

VAIL, P R, R M MITCHUM R G TODD, J M WIDMIER, S THOMPSON, J B SANGREE, J N BUBB and W C HATLELID (1977). Seismic stratigraphy and global changes of sea-level. **Memoir 26, American Association of Petroleum Geologists,** 49-212.

WALLING, D E (1984). The sediment yields of African rivers. **International Association of Hydrological Societies publication** 144, 225-239.

WATTS, A B (1982). Tectonic subsidence flexure and global changes of sea level. **Nature**, 297, 469-474.

WILLIAMS, D F, W S MOORE and R S FILLON (1981). Role of glacial Arctic Ocean ice sheets in Pleistocene oxygen isotope and sea-level records. **Earth and Planetary Science Letters,** 56, 157-166.

WORSLEY, T R and T A DAVIES (1979). Sea-level fluctuations and deep-sea sedimentation rates. **Science**, 203, 455-456.

Chapter 3

SEDIMENT DISTRIBUTION AND DYNAMICS ON THE NATAL CONTINENTAL SHELF

Burg Flemming
Senckenberg Institut, Wilhelmshaven, West Germany

Rowena Hay
Department of Geotechnology, Pretoria

INTRODUCTION

Detailed marine sedimentological investigations off Natal started less than 10 years ago. As exemplified by Siesser et al. (1974), little if any information on sediment dispersal off the entire east coast of Southern Africa existed prior to the mid-1970s. A preliminary overview of textural and compositional aspects, based on a limited number of grab samples, was presented by Moir (1976). This was followed by systematic side-scan sonar surveys, in the course of which a large variety of current-generated bedforms were discovered (Flemming, 1978, 1980a). A broadly-scaled evaluation of the sonograph data was subsequently synthesized into a regional bedload dispersal model (Flemming, 1981).

The model revealed a number of important features, especially some concerning the directions of transport of bedload material along different coastal sectors (cf. Flemming, 1981, Figure 14). Thus the east coast shelf was divided into four sedimentary compartments which were separated from each other by bedload partings, the two largest partings being situated off the Natal coast (Figure 3.1). Since tidal currents could be discounted as viable transport agents, the diverging flow patterns inferred from the different bedform migration paths had to have other causes.

As outlined by Flemming (1981), several interacting factors were needed to explain the complex structure of the dispersal patterns. The most obvious, and without doubt, most important factor is the Agulhas Current. This western boundary current hugs the continental margin

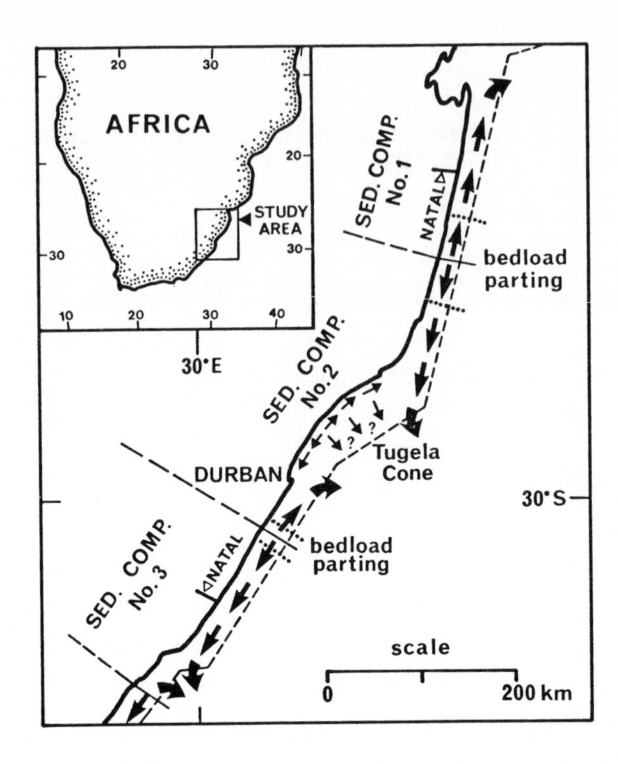

Figure 3.1 Regional setting of study area with schematic bedload dispersal model.

over large distances, and locally flows at mean velocities of over 2.5 m/s (Pearce et al. 1978; Schumann, Chapter 5). While this current could explain all the southwestward-facing sand streams and their associated bedforms, additional factors had to be found for the formation of local countercurrents in order to account for the observed bedload partings. The partings are invariably situated at the southern limits of large clockwise eddy systems, occupying embayments formed by structural offsets in the coastline. Such eddies appear to be the result of topographically induced vorticity changes in the geostrophic flow (Gill and Schumann, 1979), while shelf currents have been found to be very responsive to atmospheric forcing (Bang and Pearce, 1978; Schumann, Chapter 5). Since coastal low-pressure systems pass through the area at regular intervals (Hunter, Chapter 4), such return flows take on a semi-permanent character, thereby providing a suitable mechanism by which the formation of bedload partings and the associated northward transport of sediment can be explained.

Another important feature which emerged from the sonograph data was the dynamic behaviour of the bedload partings. They were found to shift periodically up and down the coast over certain distances. Off the northern Zululand coast, for example, the parting zone occupies a shelf section which is at least 100 km long, whereas off Scottburgh it extends for barely more than 10 km. The cause of these lateral oscillations can most probably be found in the variations in intensity and mean flow path of the Agulhas Current (e.g. Gründlingh, 1986). As a consequence of this periodic migration of the bedload parting numerous pseudo-tidal effects are superimposed on the features resulting from the otherwise unidirectional current (Flemming, 1987a). The most important of these are dunes which occasionally migrate in opposing directions, inflected megaripple crests, symmetrical megaripples, reactivation surfaces inferred from temporarily degraded dunes, and large submerged, spit-bar complexes which resemble tidal sand bars (Flemming, 1987a).

While the large-scale elements of the sediment dispersal model outlined above are well documented, knowledge of numerous aspects concerning the finer details is still inadequate. For example, the existence of bedload parting zones and their locations have been known for some years (Flemming, 1981), but little has been published on their detailed structure. Similarly, the relationship between sediment distribution and dynamics and the local wave and/or current regimes on various sectors of the shelf has not been clarified before. However, over the past few years, a great deal of new sedimentological and geophysical data has become available, especially from the Natal continental shelf (Flemming and Hay, 1984; Hay, 1984; Felhaber, 1984; Martin, 1985). The purpose of this chapter is to incorporate this new information into the existing bedload dispersal model and thereby to throw some light on the finer details referred to above.

METHODS

Mean annual sediment supply rates by local rivers to the Natal coast were calculated for individual catchment areas using a detailed sediment-yield map published by Rooseboom (1978). The map is based on numerous environmentally sensitive factors such as geology, soil cover, slope, nature and state of the vegetation cover, seasonal rainfall patterns and locally measured solids discharge of the larger rivers. This makes it by far the most reliable source of information on sediment yields available to date, being superior to the more indirect measure by Schwartz and Pullen (1966) and Midgley and Pitman (1969), which was based on the extrapolation of some locally measured sediment yields and their relationship to catchment area, mean annual rainfall and drought frequency data. A comparison of the two methods indicates that the latter approach can lead to unrealistically high estimates.

Sediment samples were recovered partly by van Veen grab (Moir, 1976; Birch, 1979) and partly by Shipek grab (Hay, 1984; Felhaber, 1984). The samples were dialyzed and washed through a 63 µm sieve to separate sands and muds. Grain-size distributions of the sand fractions were obtained by means of an automatically recording settling tube system (Flemming and Thum, 1977) and textural data were computed using moment statistics. Sediment distribution patterns were plotted and contoured by hand. Critical shear velocities (u^*_{crit}) for a water temperature of 20°C were determined from a grain size vs. shear velocity graph based on Shield's criterion for the initiation of bedload transport (Blatt _et al._ 1980). Critical surface velocities (u_{dcrit}) were calculated by substitution assuming a logarithmic velocity profile of the form

$$u_z = \frac{u^*}{k} \ln\left(\frac{z + z_o}{z_o}\right)$$

where u_z is the velocity in m/s at a height z m above the bottom (in this case $u_z = u_d$, where d is the water depth in m), k is the von Karman constant (=0.4) and z_o is the roughness length in m, here approximated by the mean grain size (D). Since $d \gg z_o$, the equation can be simplified to read

$$u_{dcrit} = 2.5\, u^*_{crit} \ln(d/D)$$

The critical surface velocities required for initiation of bedload transport and suspension were calculated by substituting the graphically determined u^*_{crit} values together with the water depth and the mean grain size of each sample locality. In this manner quasi-synoptic critical surface velocity charts were compiled for some sections of the Natal continental shelf, enabling a rapid qualitative assessment to be made of potential current influence on the seabed. It should be emphasized though, that this method predicts only unidirectional current influence and not wave or combined wave/current effects. Since wave-induced currents are superimposed on the ambient current field, the calculated critical surface velocities are more likely to give underestimates rather than overestimates of the seabed dynamics, especially in shallow coastal waters.

Bedforms and seabed physiography were mapped in the course of systematic side-scan sonar surveys, using either an E G & G Mark IB or a Klein Hydroscan 520 dual-channel system. The former instrument operates at a frequency of 105 kHz, has a beam angle of 1.2° and plots on 11-inch paper. The latter scans at a 100 kHz, has a beam angle of 1.0° and plots on 19-inch paper. Neither system produced isometric images and no attempts were made to eliminate scale distortions on the sonographs. Evaluation and interpretation of the records followed the procedures recommended by Flemming (1976, 1980b).

The shallow structure of the Natal continental shelf, in particular of Holocene sediment accumulations, were mapped by high-resolution seismic reflection profiling (see also Martin and Flemming, Chapter 2). Most of the data were collected with a multi-electrode sparker sound source, operating from a 500-joule power supply and recording over a bandpass of 400-600 Hz which achieved a vertical resolution of 3 to 5 m. Locally an E G & G boomer system, operating from a 300-joule power supply was used.

RESULTS

Sediment Supply

Potential sediment sources along the Natal coast are fluvial discharge, coastal and submarine erosion, aeolian transport, biogenic products in the form of skeletal carbonates and *in situ* authigenic mineralizations.

Of these potential sources, authigenic mineralizations and aeolian input are negligible and will not be considered further (Flemming,

1981). The annual supply of sediment from coastal and submarine erosion is difficult to assess because no large-scale erosion of modern shorelines can be observed, other than local man-induced processes (e.g. the Durban beaches). It may therefore be assumed that at present the annual amount is small, achieving significant proportions only over long time intervals, especially in the wake of positive or negative sea-level changes. In addition, there is an appreciable export of beach sands to adjacent beach ridges and coastal dunes (e.g. Weisser et al. 1982; Nicholson, 1983; Weisser and Baker, 1983), which should, in part, counterbalance any input from coastal erosion.

This leaves fluvial discharge and biogenic production as the most important short-term sources of sediment on the Natal continental shelf. A first estimate of the annual volumetric input from local rivers was made by Dingle and Scrutton (1974). They used a highly simplified approach (cf. Schwartz and Pullen, 1966; Midgley and Pitman, 1969) in which sediment yield was estimated on the basis of mean annual runoff, catchment area and drought-frequency data. For the Tugela River alone they calculated an annual sediment yield of 62×10^6 m^3. By comparison, Flemming (1981), using a different method, derived a figure of only 4.4×10^6 m^3 for the same river - a value that was subsequently modified to 5.6×10^6 m^3 (Flemming and Hay, 1983; 1984). This latter figure compares very favourably with data recently published by Nicholson (1983), who calculated an annual volumetric solids discharge from the Tugela River of 5.1 to 6.3×10^6 m^3. Bearing in mind that there are many unknowns in sediment yield calculations, overall global figures (Milliman and Mead, 1983) would suggest that the values of Dingle and Scrutton (1974) are almost certainly too high, whereas those of Flemming and Hay (1983) appear to be more realistic.

Extracting the data for total annual sediment yield for coastal Natal from the data of Flemming and Hay (1983) gives a figure of some 20×10^6 m^3 or about 30×10^6 metric tons of terrigenous sediment input per year. If it is assumed that the bedload component makes up at least 12 per cent of the total (Rooseboom, 1982), then the annual suspended load and bedload inputs by local rivers amount to 17.6×10^6 m^3 (26.4×10^6 metric tons) and 2.4×10^6 m^3 (3.6×10^6 metric tons), respectively. It should be added here that modern sediment yield values exceed long-term averages by a factor of between 12 and 30, in estimates of total land erosion (Murgatroyd, 1979) or the volumetric content of sedimentary basins around the subcontinent since the break-up of the West Gondwana landmass (Martin, 1984; Goodlad, 1986). The main reason for this discrepancy is probably the dramatic increase in soil erosion due to poor farming practices over the past 100 years or so. Caution should thus be exercised in extrapolating modern sediment yield data backwards in geological time.

The second important source of shelf sediments is the biogenic pool, comprising mainly sand- and gravel-sized bioclastic debris as reflected in calcium carbonate-contents of the surficial sediments. Flemming (1978) and Hay (1984) have shown that carbonate contents of local shelf sediments range from less than 10 per cent to greater than 90 per cent by mass, generally increasing with water depth (Figure 3.2). Since the middle and outer shelf carbonates are mainly relict in nature and the nearshore component is young, the average carbonate content of nearshore sediments may be equated with the annual production in order to achieve the observed regional trend. Flemming (1981) and Flemming and Hay (1984) used a figure of 15 per cent of the bedload component, but more recently there have been indications that this might have been an overestimate (A K Martin, pers. comm.). As a compromise and probably a reasonable approximation a value of around 5 per cent might be more realistic. This would yield an annual input of about 0.18 x 10^6 m^3 (0.27 x 10^6 metric tons) of biogenic carbonates to the whole Natal shelf. The total annual bedload supply to the Natal shelf is thus estimated at 2.58 x 10^6 m^3 or 3.87 x 10^6 metric tons (Figure 3.3).

Sediment Distribution

Sediment distribution patterns are based on textural analyses of 370 grab samples, with the areal coverage varying considerably. Coverage is generally good, and locally excellent, between Port Edward and Cape Vidal (Figures 3.4A, 3.7A), but is very limited further north. This northern shelf sector will therefore not be dealt with here. Due to the large length-to-width ratio of the shelf the textural data are presented for two geographic blocks, one covering the shelf between Cape Vidal and Durban (Figures 3.4, 3.5, 3.6), the other the shelf

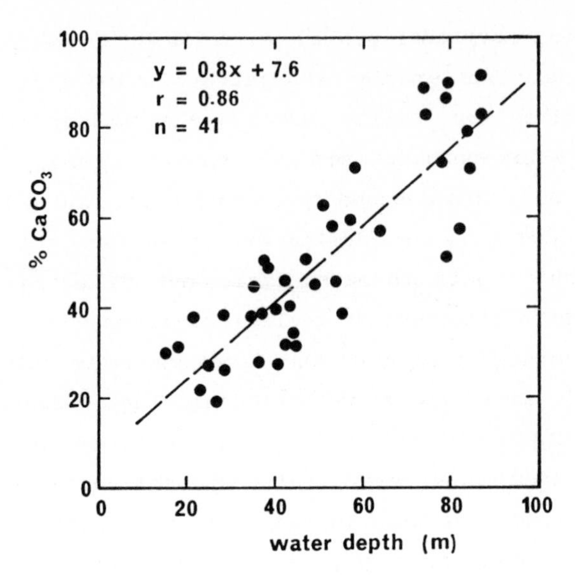

Figure 3.2 Biogenic carbonate content of shelf sediments as a function of water depth (Scottburgh-Port Shepstone).

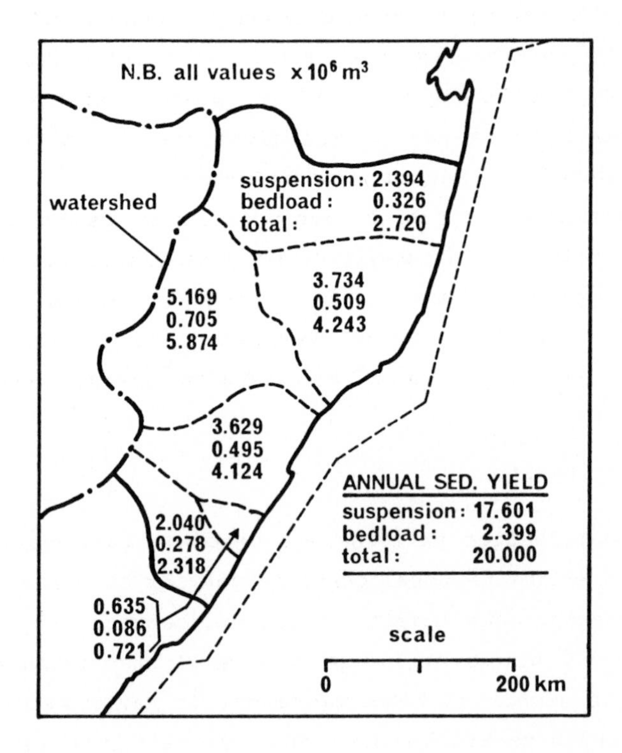

Figure 3.3 Annual sediment yields of catchment areas in Natal. (N.B. all values x 10^6 m^3).

between Durban and Port Edward (Figures 3.7, 3.8, 3.9). The percentage distribution patterns of the following parameters are presented: gravel, sand, mud, very coarse and coarse sand, medium sand, fine and very fine sand, as well as mean diameter, sorting and percentage calcium carbonate. A sample location map and a bathymetric map complete each set.

The shelf section between Durban and Cape Vidal is roughly 230 km long and 5 to 10 km wide, and the shelf break occurs at a depth of 100 m. From this area 160 sediment samples were taken for analysis, 113 alone from a 40 km x 40 km block off the Tugela River mouth (Figure 3.4A).

The surficial sediments consist predominantly of sand-sized material (0.063 to 2 mm) the content ranging from 50 to 100 per cent (Figure 3.4C). Small amounts (1 to 10 per cent) of gravel (>2 mm) are present in most samples, with nearshore sediments tending to have lower gravel contents than offshore sediments did (Figure 3.4B). Local highs with gravel values up to 60 per cent can occur on the middle shelf, always associated with partially reworked aeolianites that form an almost continuous ridge based at about -65 m, where it traces the position of a former sea-level stand. Denser and more concise sampling would without doubt reveal a similarly continuous gravel belt lining the ridge - a feature borne out by numerous sonographs (e.g. Figure 3.12). Other small areas with high gravel contents, e.g. those near the shelf break off the Tugela River, are linked to bedrock outcrops which locally penetrate the surficial sediment cover. As with the sands, the gravels comprise material from both terrigenous and biogenic sources.

True mud deposits are rare along the entire coast of southern Africa, in spite of the relatively large annual input described earlier. Notable exceptions from this regional trend are the two mud depocentres off the Tugela River and a smaller one off St Lucia (Figure 3.4D). However, even these deposits contain only minor proportions (<10 %) of the estimated local input (Felhaber, 1984), with most of the suspended sediments being exported from the region. Furthermore, as shown by Felhaber (1984), the offshore muds off the Tugela River differ from the nearshore muds in colour, degree of compaction and geochemical composition. Thus, whereas the nearshore muds are brown and soft, the

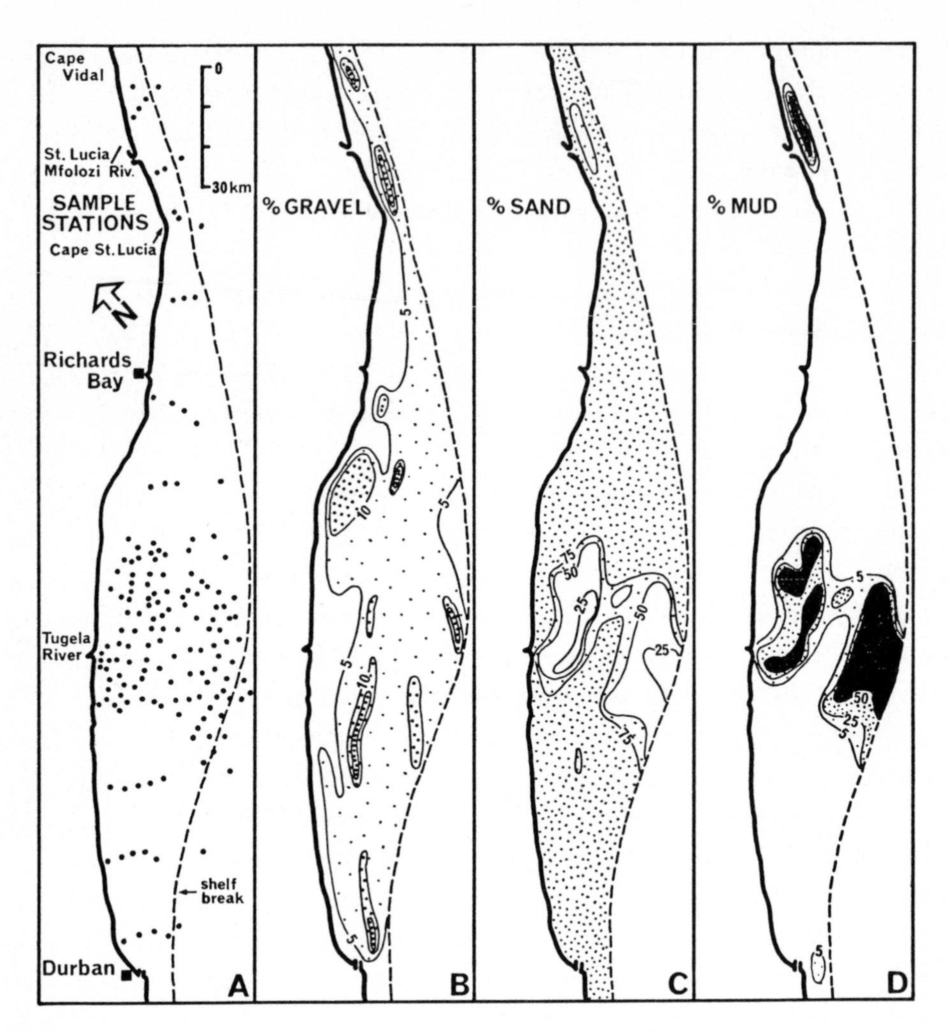

Figure 3.4 Sediment distribution patterns (northern shelf sector).
A. Location of sample stations.
B. Distribution of gravel (>2 mm).
C. Distribution of sand (0.063-2.0 mm).
D. Distribution of mud (<0.063 mm).

offshore ones are grey and stiff, with a much lower clay content. The latter have been dated at older than 40 650 years B.P. (Pta 1611), suggesting an age associated with a Pleistocene low sea-level stand (Flemming, 1978, 1980; Felhaber, 1984).

Since the bulk of the surficial sediment comprises sand (> 75 %) with gravel and mud restricted to well-defined localities, the regional dispersal patterns are better illustrated by the distribution of individual sand fractions. It is evident from Figure 3.5A that the

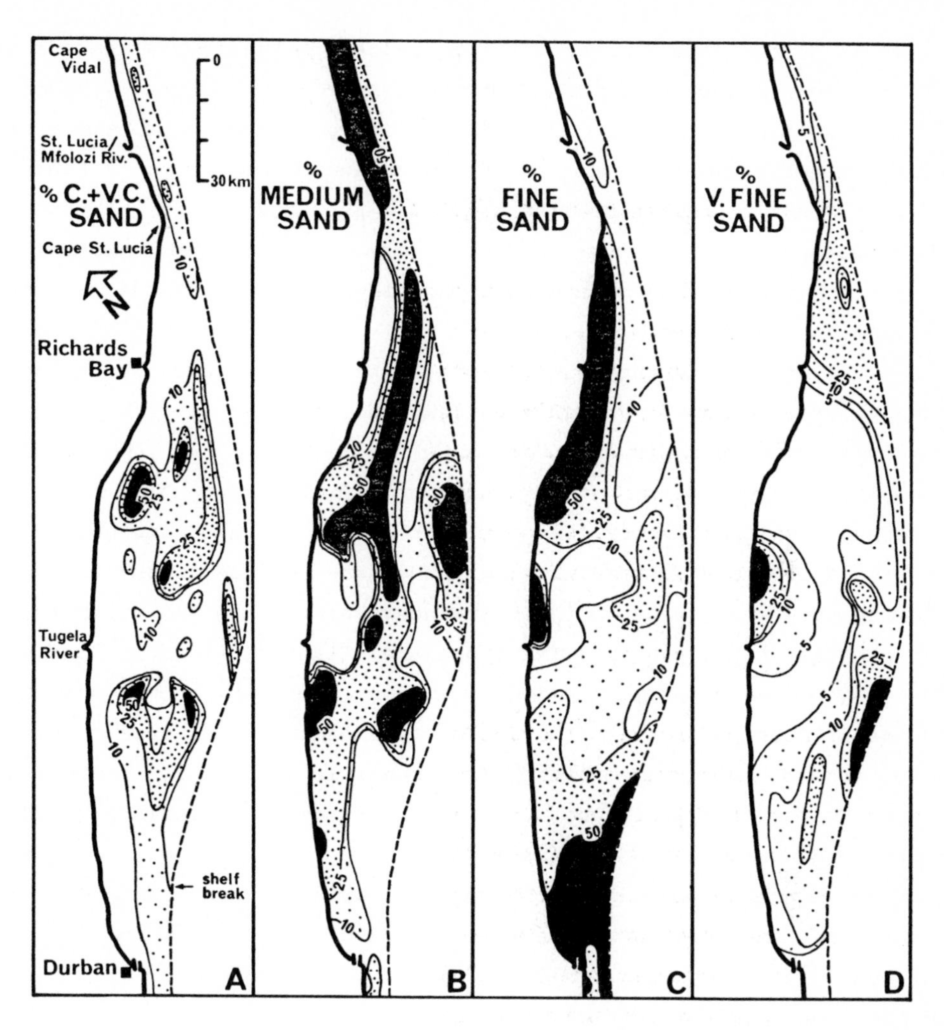

Figure 3.5 Distribution of individual sand factions (northern shelf sector).

A. Distribution of coarse and very coarse sand (0.5 - 2.0 mm).
B. Distribution of medium sand (0.25 - 0.5 mm)
C. Distribution of fine sand (0.125 - 0.25 mm)
D. Distribution of very fine sand (0.063 - 0.125 mm)

combined coarse and very coarse sand fractions follow a trend very similar to that of the gravel, being concentrated around the midshelf dune ridge and other rock outcrops on the outer shelf. Medium sand (Figure 3.5B) increases this trend further in the central and outer shelf regions of the study area, and is additionally concentrated in the nearshore between the Tugela River and Durban, as well as northwards from Cape St Lucia. This pattern suggests that medium sand has several sources, one being the midshelf dune ridge, another the Zinkwazi, Enonoti, Mvoti and Mdloti Rivers in the south and a third source the Mfolozi/St Lucia River confluence in the north. Fine sand (Figure 3.5C) in turn fills the nearshore gap between Cape St Lucia and the Tugela River and occupies most of the southern shelf section just north of Durban, with a progressive northward-declining tail bordering the shelf break.

The very fine sand fraction (Figure 3.5D) appears to be hydraulically associated with the fine sand fraction, at least in part, suggesting a common source for much of the finer sediments. In the south it completely overlaps with the distribution pattern of fine sand, the depocentre being slightly offset to the north, perhaps indicating some degree of progressive size sorting. The same applies to the region off the Tugela River. Here, both in the nearshore and in the offshore, the depocentres of the two size fractions are slightly offset. Similarly, the depocentres of very fine sand and mud are offset in the nearshore. In the offshore, however, they occupy the same general area. This may point towards different hydrodynamic controls in the two regions with regard to the size limits between the finest bedload population and the corresponding suspension population.

Off Richards Bay, very fine sand occupies a much smaller, more centrally-situated coastal sector than the fine sand fraction, but extends further offshore than the latter. The remainder of the shelf is practically devoid of very fine sand.

The trends in regional grain size distribution, expressed in terms of mean diameter and sorting of the sand fractions (0.063 - 2.0 mm), are illustrated in Figure 3.6A and B, respectively. The most widely-occurring mean grain sizes are evidently those of fine and medium sand, the latter being concentrated mainly along the midshelf dune ridge and in the northern shelf sector (Figure 3.6A). Both very coarse and very find sands do not occur in sufficiently high concentrations to be respresented in the mean grain size range. Coarse sand, on the other hand, predominates locally along the dune ridge.

The standard deviation or sorting of many marine sediments was shown mostly to reflect mixing of different hydraulic populations originating from different sources, rather than progressive size-sorting of a

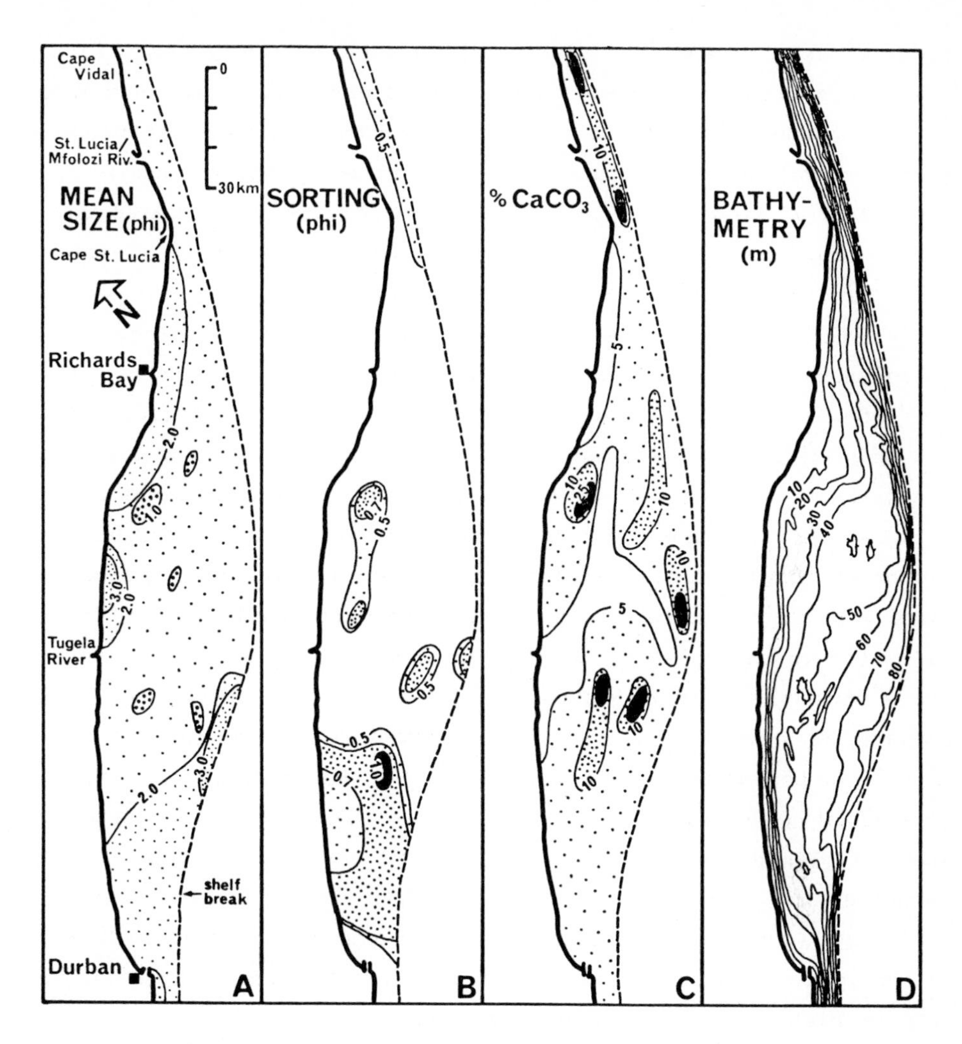

Figure 3.6 Sediment texture and composition (northern shelf sector).
A. Mean grain size in phi-values.
B. Sediment sorting in phi-values.
C. Biogenic carbonate content of shelf sediments.
D. Bathymetry of northern shelf sector.

single parent population (Flemming, 1982, 1987b). With few exceptions the same interpretation can be applied to the shelf sector under consideration here (see also Felhaber, 1984). The poorest sorting and hence, the most widespread mixing, is observed in the southern section of the shelf between Durban and the Tugela River. Indeed, it is in this area that the greatest diversity of individual size fractions can be observed (Figure 3.5), although not every size fraction also constitutes a separate hydraulic population.

The bulk of the sediments on the east coast shelf comprises a mixture of terrigenous quartz and bioclastic carbonates. In some areas the beach and nearshore sediments may contain appreciable amounts of heavy minerals. Locally, especially in the vicinity of Richards Bay, the beach sands contain such high concentrations of heavy minerals that they form exploitable placer deposits.

From Figure 3.6 it can be seen that carbonate contents generally increase with water depth, following trends similar to those in Figure 3.2. Concentrations locally reach over 50 per cent, particularly along the midshelf dune ridge and in the vicinity of rock outcrops on the outer shelf. As pointed out by Flemming (1980) the carbonates comprise a relict lag component found mainly on the outer shelf and a modern component resulting from skeletal breakdown of benthic organisms on the inner shelf. Since there is a continuous supply of terrigenous sediments to the coast by local rivers, the in situ-produced carbonates are constantly being diluted, as a result of which the shoreward decrease in carbonate contents is observed along the entire continental margin. Lowest carbonate contents occur in areas of mud deposition. This highlights the terrigenous origin of the muds. Pelagic sources seem to play a subordinate role in shelf waters, in contrast to the situation in the open ocean (Martin, 1984; Goodlad, 1986).

Finally, Figure 3.6D illustrates the strong topographic control by the Tugela Cone on the bathymetry of this northern shelf sector. Whereas off Durban and to the north of Cape St Lucia the width of the shelf rarely exceeds 10 km with a relatively well-defined break at -100 m, it reaches 40 km off the Tugela River, where the shelf break is rather ill-defined at about 100 m.

In contrast to the northern shelf, the 160 km shelf section between Durban and Port Edward is far more uniform, widening gradually from 7.5 km off Durban to 12 km off Hibberdene, from where it once more narrows to 9 km off Port Edward. Some 200 sample stations were occupied in this region, of which 160 were successful, the remainder indicating rocky bottom (Figure 3.7A).

As in the northern shelf sector, sand is by far the most dominant textural component (Figure 3.7C), reaching 100% close to the shore along the entire coastline and more than 75% over large parts of the

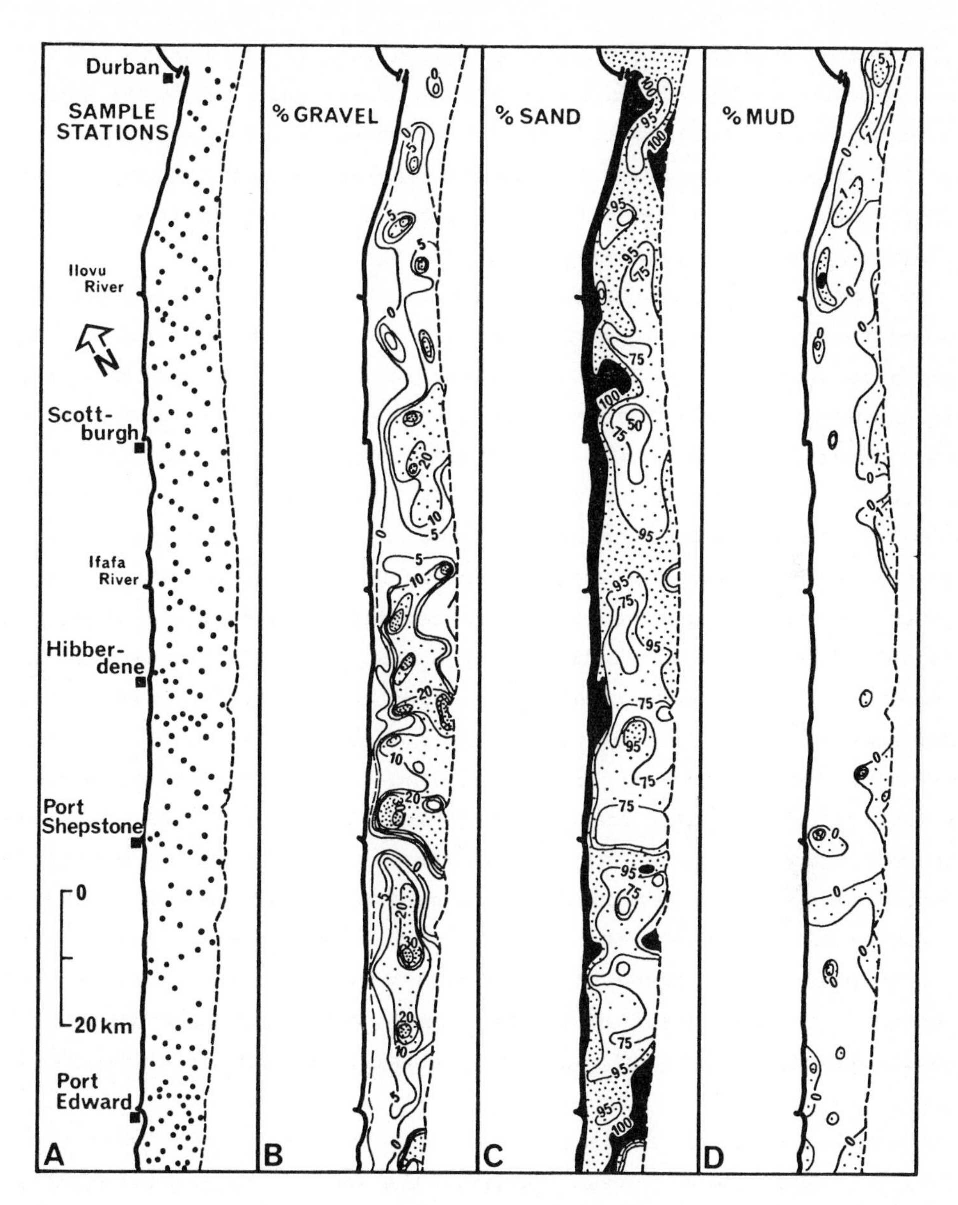

Figure 3.7 Sediment distribution patterns (southern shelf sector)
A. Location of sample stations.
B. Distribution of gravel (>2 mm).
C. Distribution of sand (0.063 - 2.0 mm).
D. Distribution of mud (<0.063 mm).

remaining shelf. A significant difference between the northern and southern shelf sectors, however, is observed in the occurrence of both gravel and mud in the southern sector. Gravel (Figure 3.7B) is far more common on the shelf south of Durban than to the north. It occurs in the form of coarse lag deposits on the middle and outer shelf in areas of erosion or non-deposition. The reverse applies to mud (Figure 3.7D), traces of which may be found in many places, but there is only a single small depocentre; this is off the Ilovu River in the lee of the large spit-bar complex (Flemming and Martin, 1985).

By tracing the distribution of individual grain size fractions, it is found that most of the combined coarse and very coarse sand fractions are closely linked to the distribution of gravel (Figure 3.8A). This would suggest that they form part of the coarse lag material, although some degree of localized size sorting may be inferred from slight shifts in the relative positions of the individual depocentres. Medium sand in turn is partitioned between two major sources (Figure 3.8B). A smaller portion appears to be genetically related to the coarse lag material, confirming that local down-current size-sorting occurs. This is particularly evident between Hibberdene and Ifafa, where all three fractions are grouped together, each with its own well-defined depocentre.

The second major source of medium sand is derived from at least two groups of terrestrial drainage systems. One comprises the rivers between Scottburgh and Ifafa, the other the rivers in the vicinity of Port Edward. In both cases large medium-grained sand sheets extend from the coast out onto the shelf. Significantly, off Scottburgh the sheet takes on the shape of a northeastwards-facing tongue, oriented obliquely to the coastline. This sand body provides indirect, but conclusive, evidence of sustained counter-current activity in this region.

A similar trend is evident in the distribution of fine sand (Figure 3.8C). Again an oblique, northeastward-facing tongue is developed, this time emanating from the coastline between Scottburgh and Ilovu. With the exception of small depocentres off Hibberdene and Port Shepstone, the plume off Ilovu forms the only major fine sand deposit on the shelf between Durban and Port Edward. Contrary to the northern shelf sector, very fine sand (Figure 3.8D) is found in traces only.

From a sediment dynamics point of view the two parallel sand tongues between Ifafa and Ilovu are excellent indicators of local near-bottom circulation patterns. Quite clearly, in this case, the two adjacent

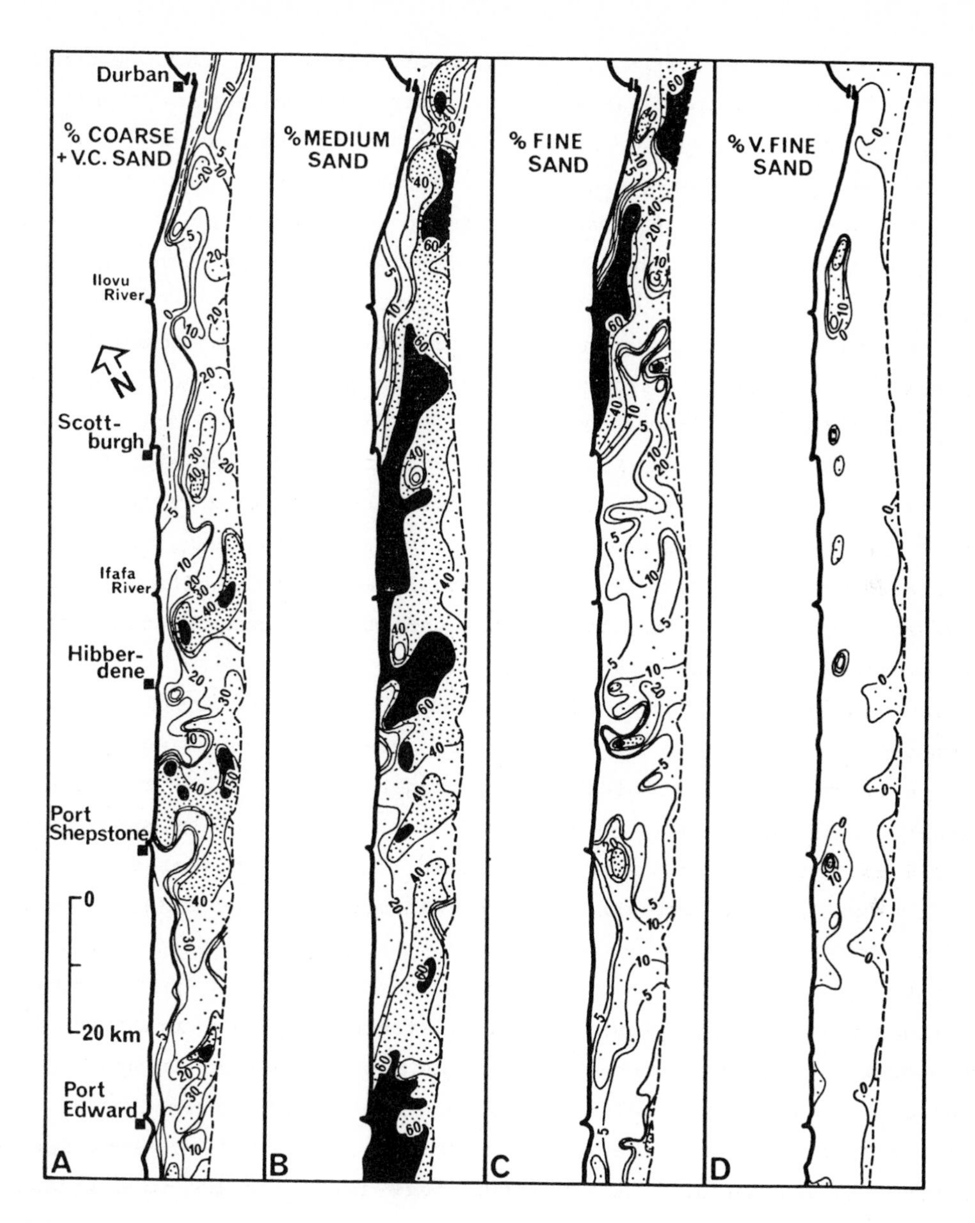

Figure 3.8 Distribution of individual sand fractions (southern shelf sector).

A. Distribution of coarse and very coarse sand (0.5 - 2.0 mm).
B. Distribution of medium sand (0.25 - 0.5 mm).
C. Distribution of fine sand (0.125 - 0.25 mm).
D. Distribution of very fine sand (0.063 - 0.125 mm).

provinces are not formed by progressive down-current size-sorting of a single parent population, but rather represent two distinct hydraulic populations supplied from different source areas. The fine sand is derived from catchment areas that drain predominantly fine-grained rocks of the Karroo Supergroup, whereas the medium sands have their source in the coarser-grained rocks of the Cape Supergroup and older granite/gneiss complexes found close to the shoreline. There are therefore two independent, but parallel sandstreams with only minor overlap.

The mean grain size pattern (Figure 3.9A) simply summarizes the overall trends observed in the distribution patterns of the individual size fractions. Sorting (Figure 3.9B), on the other hand, shows a strong inverse correlation with fine sand, i.e. areas of high fine sand concentration invariably show the lowest degree of sorting. Since better-sorted medium sands are always in close proximity to fine sand deposits, it must be concluded that the poorer sorting of the latter is caused by lateral entrainment of more medium sand into the fine-grained sand stream than *vice versa*.

The composition of the bulk sediment along this shelf sector is dominated by quartz on the inner shelf, reaching over 80 per cent very close to the shore, and bioclastic carbonates on the outer shelf, where they locally attain over 90 per cent (Figure 3.9C). The same general trend as that on the shelf to the north of Durban is therefore found, namely terrigenous domination in the nearshore and carbonate domination in the offshore.

Topographically the shelf sector between Durban and Port Edward is dominated by two major features, both revealed by major departures from a regularly sloping seabed (Figure 3.9D). The first feature is the convex, arcuate shoal between Ilovu and Scottburgh, formed by a large sand body previously referred to as the Ilovu spit-bar. The second feature is a narrow, intermittent ridge running along the midshelf subparallel to the coastline. The full extend of the ridge is not represented on the highly-simplified bathymetric chart (Figure 3.9D), but the Aliwal Shoals and Protea Banks form elevated parts of the ridge.

Sediment Accumulation

The sediment distribution patterns presented in the previous section

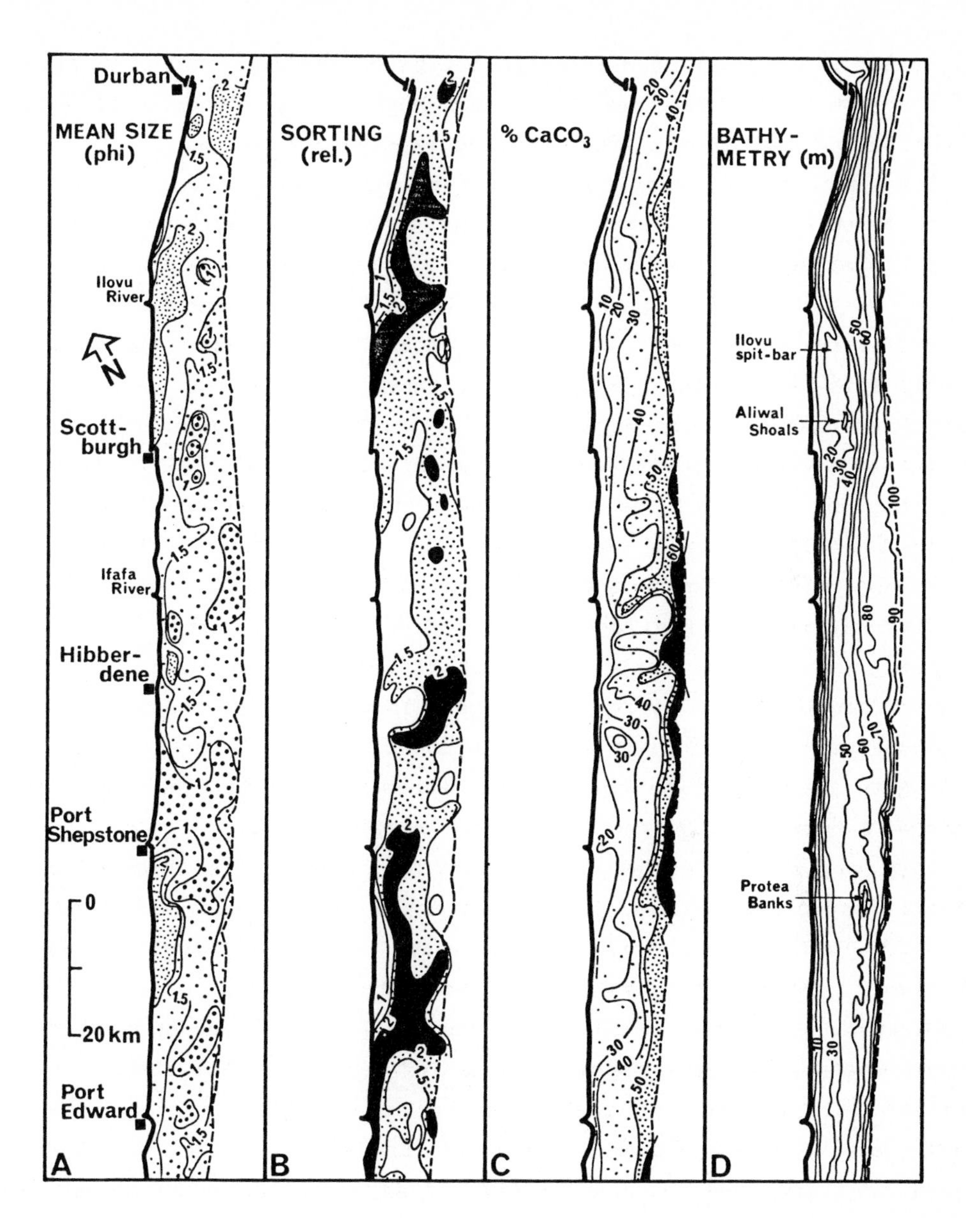

Figure 3.9 Sediment texture and composition (southern shelf sector).
A. Mean grain size in phi-values.
B. Sediment sorting in relative values.
C. Biogenic carbonate content of shelf sediments.
D. Bathymetry of southern shelf sector.

have yielded a great deal of information on sediment sources, transport routes and the type of sediment likely to occur off Natal. They do not, however, provide any data on local sediment thickness and in order to identify such sediment bodies on the shelf, shallow seismic reflection techniques have been employed. The results of such a regional seismic survey are presented in Figure 3.10A and B.

The isopachyte maps reveal that sediment thicknesses vary considerably from place to place and that major sediment bodies are very localized. Thus a major deposit, reaching thicknesses of over 30 m, is situated off the Mfolozi/St Lucia River confluence in the very north of the study area. A second sediment body, which includes the St Lucia spit-bar complex, occupies the inner shelf between Cape St Lucia and Richards Bay. It also reaches a thickness of over 30 m locally. There is a third deposit, up to 20 m thick, in a midshelf position on the shoreward side of the dune ridge to the southeast of Richards Bay. This is followed by a continuous belt of sediment which generally exceeds 5 m in thickness and which begins in a midshelf position at the northern limits of the Tugela Cone, from where it curves towards the shore to occupy the entire inner shelf between the Tugela River and Port Shepstone. North of Durban individual depocentres attain thicknesses of just over 10 m, whereas south of Durban two major deposits are found with thicknesses exceeding 30 m off the Mlazi River and 40 m between Ilovu and Scottburgh, respectively.

Only two of the larger sedimentary deposits can be recognized by their topographic relief. These are the two spit-bar complexes off Cape St Lucia and Ilovu, respectively. Both sediment bodies are actively prograding parallel to the shore across other Holocene sediments, the former to the north and the latter to the south. The St Lucia spit-bar encompasses about 2 km^3 of sediment. All other deposits either fill depressions on the shelf or are incorporated into the nearshore sediment wedge. As they form part of the normal shelf gradient, they are topographically not as conspicuous, although they can contain considerable amounts of sediment. Thus both the deposit off the Mfolozi/St Lucia River confluence and the one off the Mlazi River each contain over 1 km^3 of sediment.

Significantly, most of the Holocene sediment accumulations are concentrated within the first 5 to 10 km of the shoreline, the only exception being that in the northern Tugela Cone area. However, even there the midshelf dune ridge evidently forms the seaward boundary to cross-shelf dispersal. This confirms the semi-continuous nature of

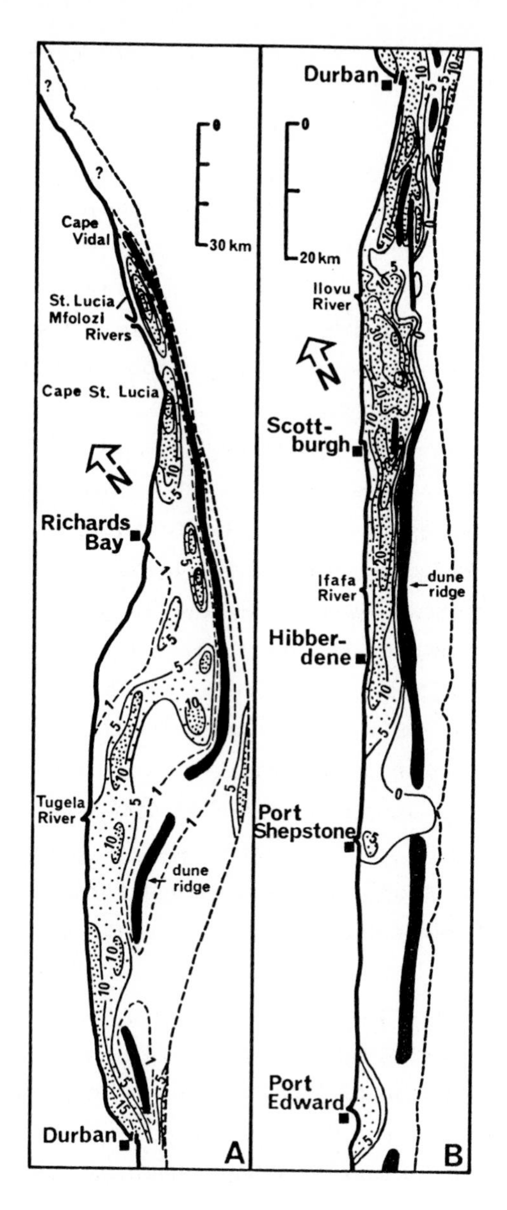

Figure 3.10 Holocene thicknesses (in m) on the Natal continental shelf.
A - northern shelf sector
B - southern shelf sector

the ridge. In fact, between Ilovu and Hibberdene the ridge acts as an effective dam behind which sediments have accreted vertically to the point of overtopping in some places. This barrier effect is well illustrated in seismic sections by Flemming (1981) and by Martin and Flemming (Chapter 2). Only off Durban and in the northern Tugela cone region is there a relatively thick deposit (> 10 m) along the shelf break. In all other offshore regions local sand sheets do not generally exceed the 1 m resolution limit of the seismic sound source.

Bedform Patterns

As has been shown by Flemming (1978, 1980, 1981), large areas of the southeast African continental shelf are sculptured by the vigorous action of the Agulhas Current and large ocean swells. With respect to bedload dispersal the shelf off Natal can be divided roughly into three major sectors, two of which are predominantly current-controlled, with the third one coming mainly under wave influence.

The first current-controlled shelf sector incorporates the entire northern shelf from the Mozambique border down to Cape St Lucia. From there southwards it no longer encompasses the whole shelf width, but hugs the outer shelf beyond the midshelf dune ridge until it is lost at the shelf break to the south of Richards Bay. This whole area is dominated by current-generated bedforms, especially transverse ones (Figure 3.11A). To the north of Cape Vidal the bedforms migrate in a northeasterly direction, whereas south of the cape they face towards the southwest, thus defining a bedload parting in the vicinity of Cape Vidal. In reality, the bedload parting occupies an almost 100-km wide corridor in which bedforms periodically switch direction (Flemming 1981, 1987a). Other bedforms include sand ribbons and sand streamers as well as a variety of longitudinal bedforms such as erosional tails and furrows.

Since the northern Natal shelf never exceeds 5 km in width, the sandstreams defined by the bedform associations occupy a 3 to 4 km-wide belt. This trend is maintained to the south of Cape St Lucia where the shoreline swings towards the southwest, whereas the sandstream continues to follow the shelf break until it spills over onto the upper slope of the Tugela Cone where the shelf break in turn changes direction.

To the south of the current-controlled northern shelf sector, the wave-dominated sector occupies most of the Tugela Cone region between Richards Bay and Durban (Figure 3.11A), with the exception of the narrow belt along the northeastern margin discussed above. The seabed physiography of this area is markedly different from the current-controlled regions with their characteristic bedform associations.

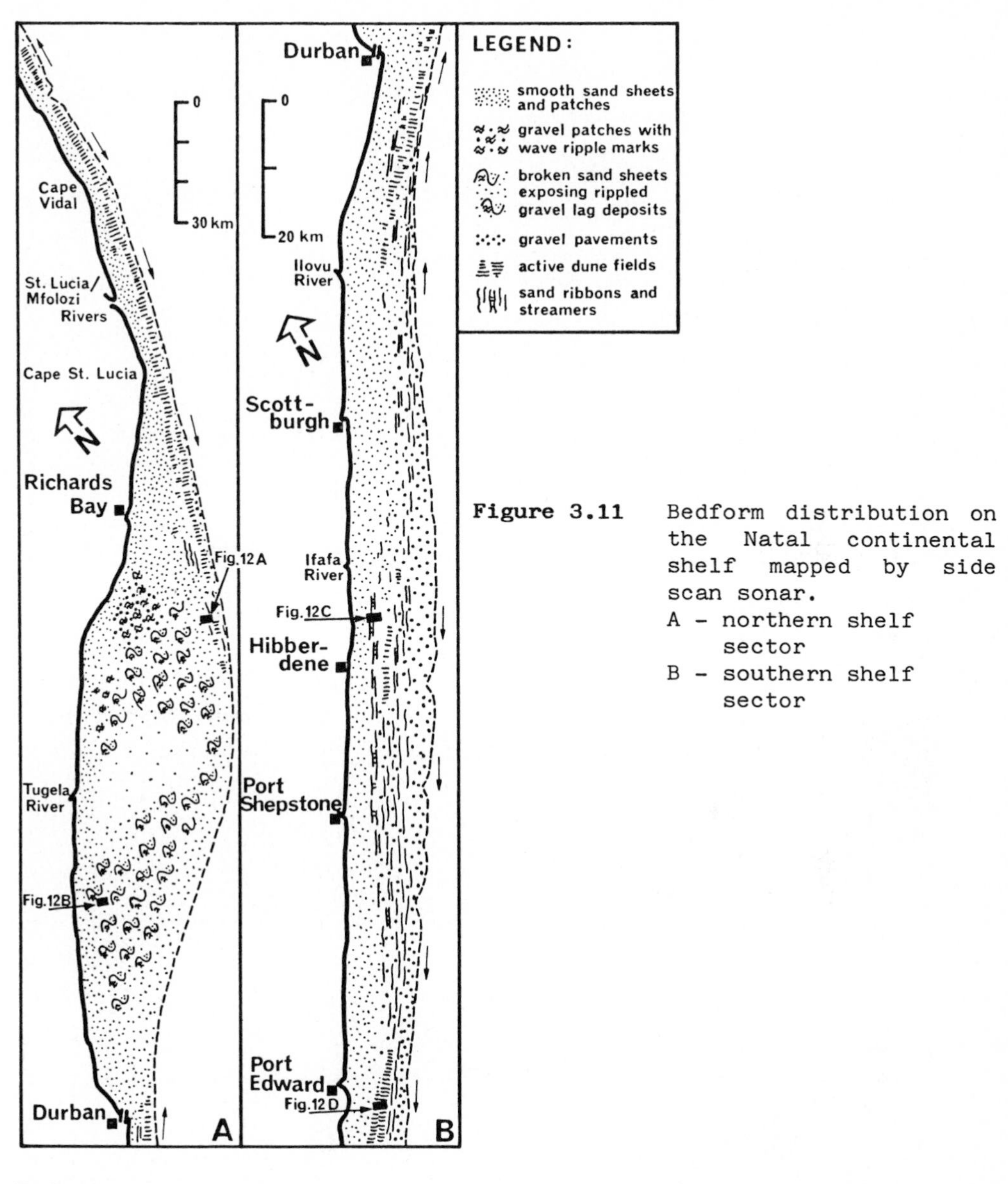

Figure 3.11 Bedform distribution on the Natal continental shelf mapped by side scan sonar.
A - northern shelf sector
B - southern shelf sector

Indeed, the whole area is devoid of any bedforms generated by unidirectional currents. Instead, large continuous tracts covered with sharp-edged sand patches occur, often displaying lobate shapes. The areas between the sand patches are occupied by rippled sediments. The spacing of the ripples and the acoustic intensity of the rippled surfaces suggest coarse sand and gravel. The sharpness of both the ripple marks and the sand patch boundaries suggest that both features are formed contemporaneously in response to vigorous wave-induced

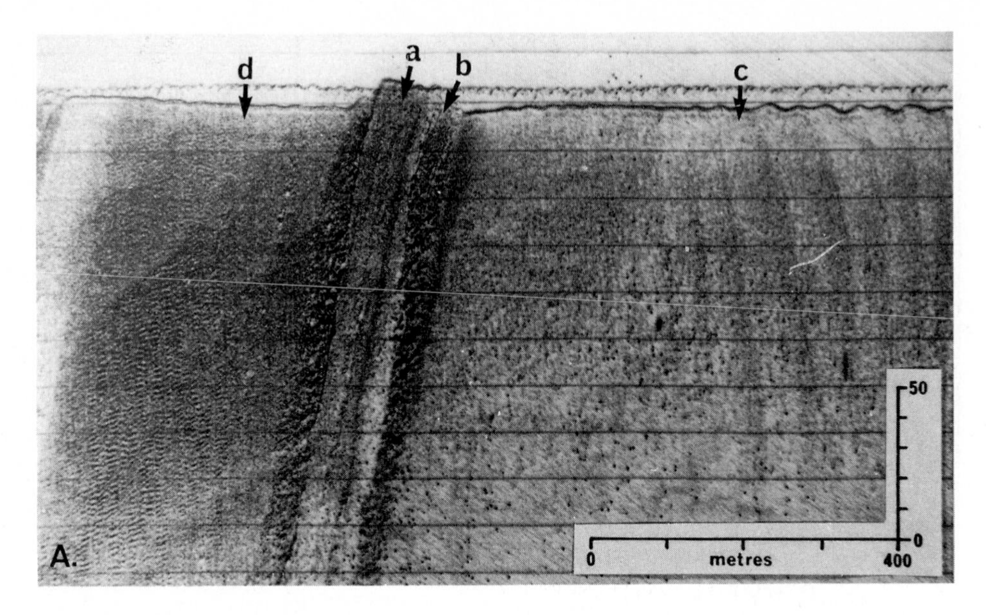

A. Pleistocene coastal dune ridge south of Richards Bay.

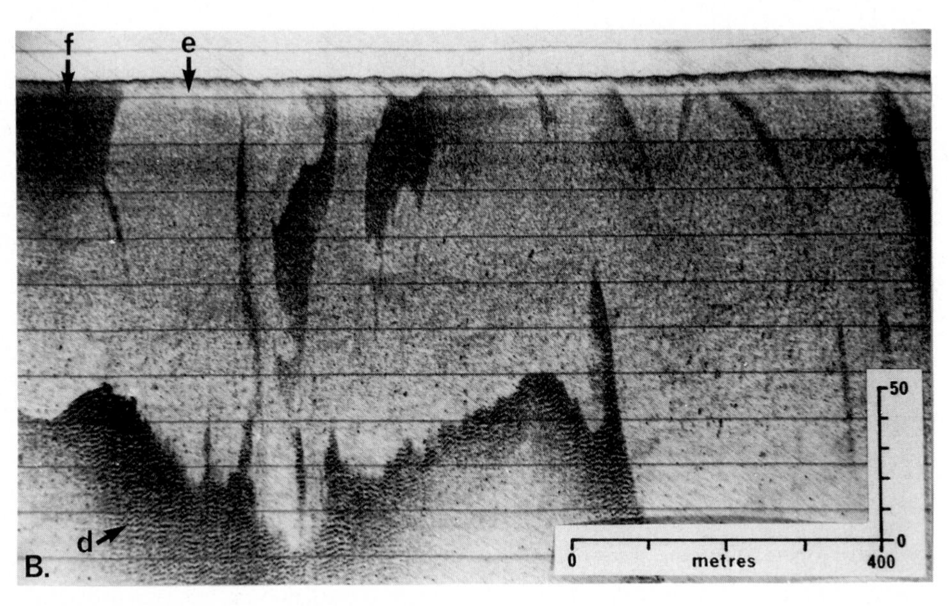

B. Wave-induced bedform patterns south of the Tugela River.

Figure 3.12 Sonographs of recurring features on the Natal continental shelf (locations indicated in Figure 3.11). Legend: a - coastal dune ridge, b - rubble slope, c - sand sheet with dunes, d - rippled gravel, e - smooth sand sheets, f - smooth gravel beds, g - sand ribbons.

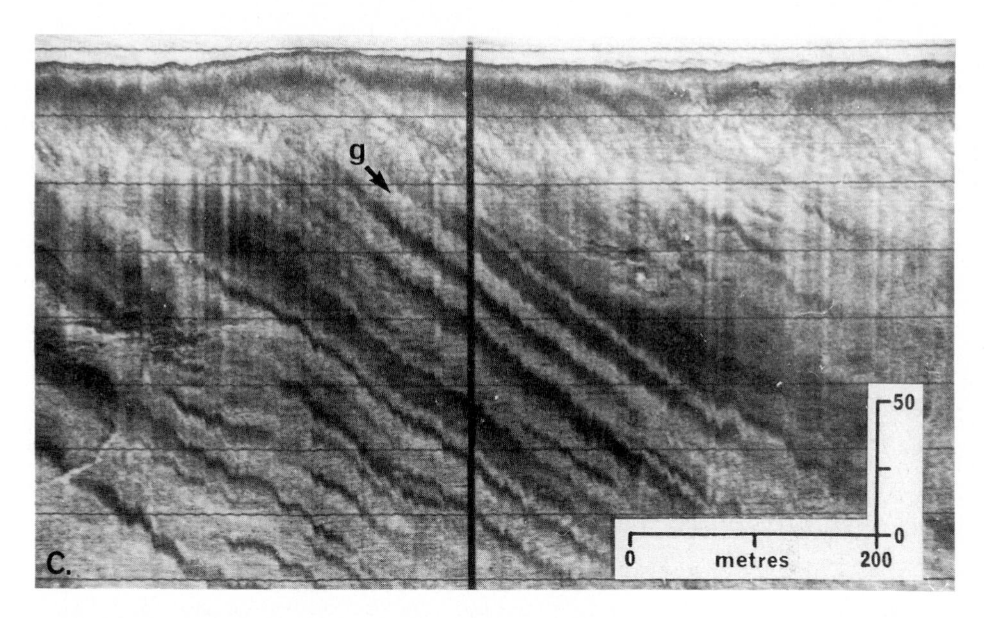

C. Sand ribbons south of the Ifafa River.

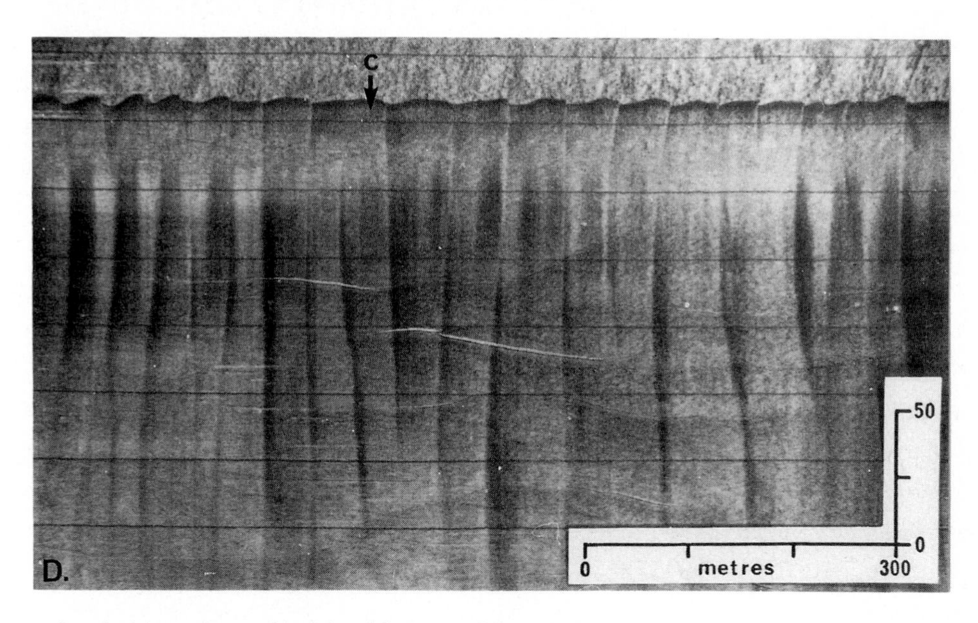

D. Active dune field off Port Edward.

turbulence and oscillatory flows at the seabed. Such ripple marks in gravelly sediments with wavelengths of up to 2 m have been found as deep as the shelf break, i.e. at 100 m water depth. An acoustically smooth and featureless zone occurs some distance offshore due west of the Tugela River mouth, coinciding with the area texturally mapped as mud. The fact that a mud depocentre has been established in a shelf environment affected by strong wave action would suggest that the muds are supplied in pulses of high concentration during seasonal river floods, and that the suspended plume becomes trapped in an eddy centred over the mud deposits.

The shelf sector to the south of Durban is once more dominated by unidirectional currents. It has been previously described in some detail by Flemming (1980) and thus only the main features will be repeated here. As with the northern sector, the shelf is characterized by two diverging sandstreams separated by a bedload parting zone. More detailed work (Hay, 1984; Flemming, 1987a) has shown this parting zone to be situated between Scottburgh and Port Shepstone, where it defines a 10 to 20 km-wide corridor that runs obliquely across the shelf. In contrast to the northern shelf sector the sandstreams do not, in this case, occupy the whole shelf, but are restricted to belts situated in a midshelf position. The nearshore zone is characterized by relatively smooth sand sheets and sediment wedges, whereas the outer shelf consists of a sand-depleted gravel pavement. Four selected sonographs illustrating typical features are presented in Figure 3.12. Other examples can be found in Flemming (1978, 1980, 1984).

DISCUSSION

Since the discovery of widespread current-generated bedforms some eight years ago, the southeast African continental margin has become a generally accepted prototype model for a shelf environment on which sedimentary processes are controlled by an intruding ocean current (Walker, 1979; Harms et al. 1982; McCave, 1985). Indeed, both the regional scale and the local dimensions of bedform development are comparable to the largest examples known to occur in shallow tidal seas. Within this concept the continental shelf off Natal plays a central role, not only because it incorporates all the important elements of the model, but also because many of them are better developed there than are those along other parts of the margin. Thus the Natal coastline straddles three of the four sedimentary compartments identified by Flemming (1980, 1981) and hence contains two of the three bedload partings separating adjacent compartments.

Although the Agulhas Current is by far the most powerful individual force at play, it is by no means the only factor involved. As demonstrated by Flemming (1981), and confirmed by the results presented here, ocean swells and wind stress currents, acting in conjunction with local topography, play important roles in the conceptual framework. After the outline of the major sediment distribution patterns given in the previous section, the discussion will focus on sediment dynamic aspects, i.e. the interaction between the sediments and the hydrodynamic dispersal agents.

The first important observation regarding such interaction concerns the general paucity of major mud deposits. Almost 90 per cent of the annual sediment input comprises suspended sediments of fluvial origin. The input rate over the past 100 years or so estimated at some 18 x 10^6 m^3 per year, together with at least 1 x 10^6 m^3 per year for the remainder of the Holocene epoch, gives a total figure of some 12 x 10^9 m^3. This would be enough to cover the whole Natal shelf with a 2 m thick mud blanket. Instead, only two relatively small depocentres share a small fraction of this area between them. Most of the fine sediments thus remain in suspension and are transported off to some remote, probably deep-sea, depository. In this respect the Natal shelf does not conform to the widely accepted contention that during interglacial periods most of the sediments supplied to the continental shelves of the world remained on the shelves (Gibbs, 1981).

In the present case the severe wave climate, coupled with wind stress and ocean currents, evidently keeps most of the muds in permanent suspension, in which form they are exported from the shelf environment. Only in places where concentrated muds get trapped in the low-velocity centres of closed eddy systems can they settle out. It is thus postulated that each of the mud deposits described is situated at the centre of such an eddy. Even if such closed cells are not permanent features, the limited extent of the mud deposits suggests that such cells must occur frequently and at preferred localities.

By contrast, all the bedload sediments must have at least an interim "shelf life" before they can be exported. Flemming (1980, 1981) showed that such bedload export does actually occur in a number of places, e.g. where currents overshoot the shelf break, thus funnelling associated sand streams on to the upper continental slope. On the Natal shelf this is observed in at least two places, one due east of Durban, where the oblique counter-current crosses the shelf break. This area coincides with a sand depocentre 10 m thick. A similar situation is found at the northeastern flank of the Tugela Cone, where the shelf break abruptly changes direction. Here too the sediment spill-over is documented by an acoustically transparent sediment body.

Bedload sediments are thus either gradually transported towards such final exit points, or they are retained in semi-permanent shelf depositories such as nearshore sediment wedges and spit-bar complexes, or in onshore depositories such as coastal dune belts and beach ridge systems. The transport routes of bedload sediments can be accurately determined in all situations where currents are strong enough to generate bedforms. Such reconstructed flow patterns are presented in Figure 3.13A and B. Good bedform data has been obtained for the northern shelf sector down to the Tugela Cone exit point (Figure 3.13A) and for the entire southern shelf sector between Durban and Port Edward (Figure 3.13B). These transport routes clearly highlight the bedload partings north of Cape Vidal and east of Scottburgh. Also outlined is a gently meandering flow mode of the Agulhas Current. At first sight this may seem surprising because much more pronounced changes in the flow path of the current have been observed in the past (Gründlingh, 1986; Schumann, Chapter 5). The fact that the bedload patterns do not reflect such short-term migrations of the current would suggest that they represent an accurate picture of the long-term mean flow path of the current - a feature that should be of considerable interest to oceanographers.

On the shelf section between Richards Bay and Durban the picture is completely different. This region is characterized by eddies (Bang and Pearce, 1978) and the absence of any current-generated bedforms would suggest that the eddy currents are rarely, if ever, strong enough to transport bedload on a significant scale. This is, in fact, confirmed by the widespread occurrence of wave-generated bedforms (Figures 3.11 and 3.12), which, in addition, do not show any evidence of current overprint.

In view of the presence of a well-defined mud depocentre off the

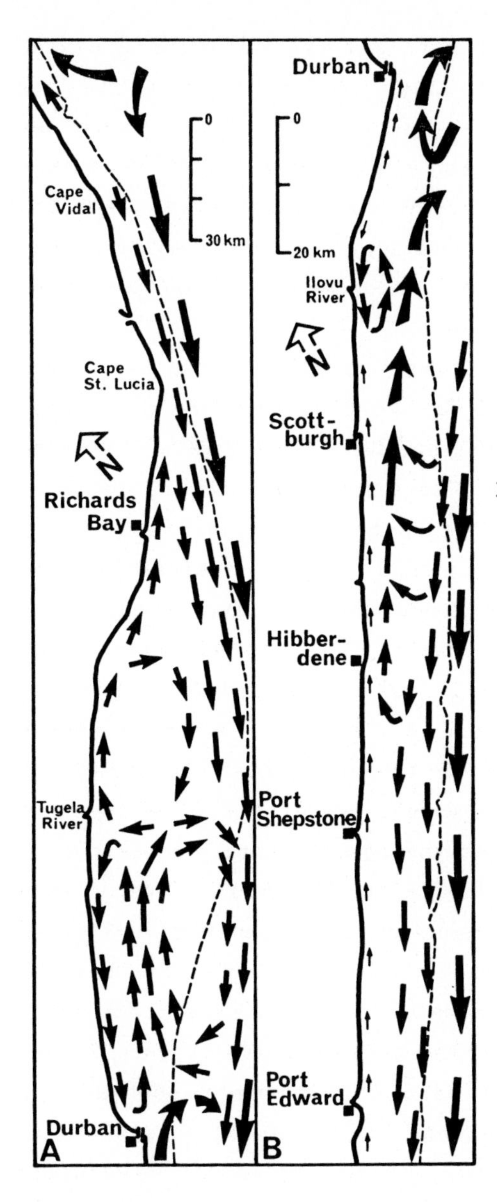

Figure 3.13 Bottom current patterns on the Natal continental shelf inferred from sediment dispersal and bedform patterns.
A - northern shelf sector
B - southern shelf sector.

Tugela River and the Mfolozi/St Lucia River confluence, it can be postulated with some confidence that these areas are frequently occupied by closed eddy systems at the centre of which mud deposition can occur. A third, much smaller eddy must be situated just north of the Ilovu spit-bar complex (Figure 3.13B).

The results presented in Chapter 5 support the possible presence of an eddy off the Tugela River, and, according to Orme (1973), the longshore drift pattern between Durban and Richards Bay shows a divide or parting just south of the Tugela River. As a result the net nearshore bedload transport is directed towards the south between the Tugela River and Durban and to the north between the Tugela River and Richards Bay. Qualitatively this pattern is in agreement with the general pattern discussed above. In summary, it can be stated that whereas bedload transport paths are well documented, the dispersal routes for suspended sediments are not.

Only on the southern shelf sector were sampling densities large enough to enable threshold conditions for both bedload and suspension transport to be modelled. The modelling was restricted to the sand fraction which was shown to represent several hydraulic populations. The widespread gravel pavements on the outer shelf were previously shown to be lag deposits (Flemming, 1980, 1981) indicating that threshold conditions would be attained only under exceptional conditions. The basic procedure for modelling sand thresholds has been presented in the section on methods. It should be added that the surface velocities predicted for threshold conditions assume that the logarithmic velocity profile extends to the surface. This is, of course, a great simplification of the actual prevailing conditions one can expect at any particular time, but the results should nevertheless provide a good approximation of the critical surface velocities required for initiation of bedload transport or suspension. Figures 3.14A and B represent synoptic threshold velocity charts for bedload and suspension respectively. In general, the predicted cross-shelf velocity levels (0.5 m/s) indicated on the bedload chart (Figure 3.14A) are of the same order as measured values (Schumann, Chapter 5). Furthermore, the gradient from relatively low threshold conditions in the nearshore to relatively high ones in the offshore corresponds closely to the offshore velocity gradient of the Agulhas Current. This would suggest that the sand-sized bedload material is by and large in equilibrium with the current regime.

In the case of critical suspension velocities (Figure 3.14B) the predicted values are close to the maximum velocities expected for most parts of the shelf. Suspension thresholds should thus be exceeded regularly on the outer shelf only. A similar conclusion was reached by Flemming (1984) for parts of the northern sector southeast of Richards

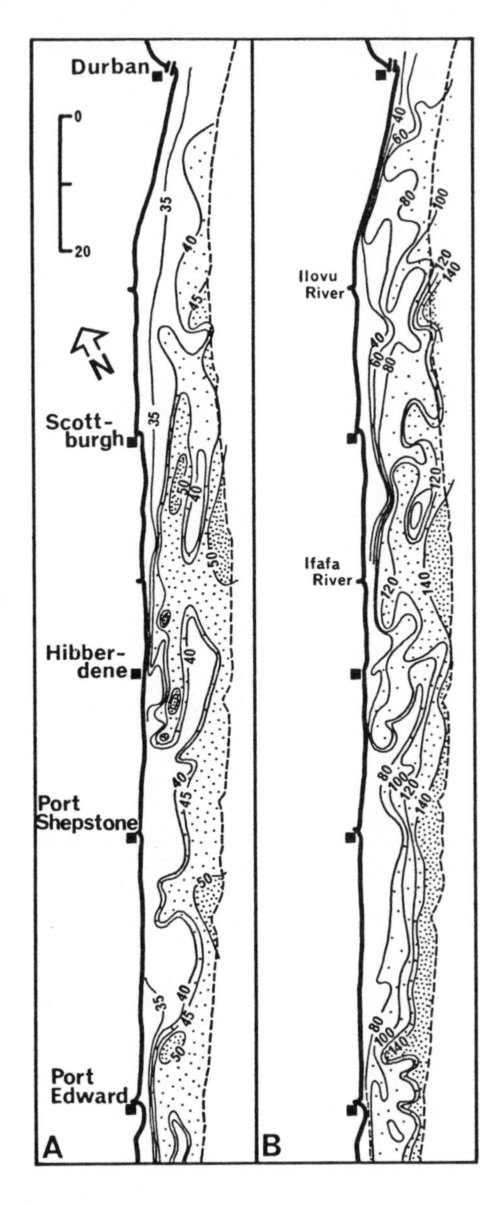

Figure 3.14 Southern shelf sector. Synoptic charts of surface threshold conditions in cm/s required for:
A - the initiation of bedload transport
B - suspension transport

Bay. This would explain the widespread existence of sand-depleted gravel pavements on the outer shelf of the study area.

REFERENCES

BANG, N D and A F PEARCE (1978). Physical oceanography. In: Ecology of the Agulhas Current Region, (Ed: A E F HEYDORN). **Transactions of the Royal Society of South Africa,** 43, 156-162.

BIRCH, G F (1979). Cruise report, R/V "Thomas B Davie" : Cruise 392. **Joint Geological Survey/University of Cape Town Marine Geological Programme** (1-20 May, 1979). 20 pp.

BIRCH, G F (1986). Unconsolidated sediments on the eastern margin of South Africa (Cape Padrone to Cape Vidal). **Bulletin of the Geological Society of South Africa,** (in press).

BLATT, H, G V MIDDLETON and R C MURRAY (1980). **Origin of sedimentary rocks.** Prentice-Hall, Englewood Cliffs, 2nd edition, 782 pp.

DINGLE, R V and R SCRUTTON (1974). Continental breakup and the development of post-Palaeozoic sedimentary basins around southern Africa. **Bulletin of the Geological Society of America** 85, 1467-1474.

FELHABER, T A (1984). The geochemistry and sedimentology of Quaternary shelf sediments off the Tugela River, Natal, South Africa. MSc thesis (unpublished), Department of Geochemistry, University of Cape Town, South Africa. 237 pp.

FLEMMING, B W (1976). Side scan sonar : a practical guide. **International Hydrographic Review,** 53, 65-92.

FLEMMING, B W (1978). Underwater sand dunes along the southeast African Continental margin - observations and implications. **Marine Geology,** 26, 177-198.

FLEMMING, B W (1980a). Sand transport and bedform patterns on the continental shelf between Durban and Port Elizabeth (Southeast African Continental Margin). **Sedimentary Geology,** 26, 179-205.

FLEMMING, B W (1980b). Causes and effects of sonograph distortion and some graphical methods for their manual correction. In: **Recent developments in side-scan sonar techniques.** (Ed: W G A RUSSELL-CARGILL). Central Acoustics Laboratory, University of Cape Town, South Africa, 103-141.

FLEMMING, B W (1981). Factors controlling shelf sediment dispersal along the southeast African Continental Margin. **Marine Geology,** 42, 259-277.

FLEMMING, B W (1982). Sediment mixing: its natural occurrence and textural expression. Eleventh International Congress on Sedimentology (IAS), Hamilton, Ontario (22-27 August, 1982). Abstracts, p. 81.

FLEMMING, B W (1984). Giant comet marks. **Geo-Marine Letters,** 4, 113-115.

FLEMMING, B W (1987a). Pseudo-tidal sedimentation in a non-tidal shelf environment. In: **Tide-Influenced Sedimentary Environments and Facies.** (Eds: P L DE BOER, A VAN GELDER AND S D NIO). Reidel, Dordrecht (in press).

FLEMMING, B W (1987b). Process and pattern of sediment mixing in a microtidal coastal lagoon along the west coast of Southern Africa. In: **Tide-Influenced Sedimentary Environments and Facies.** (Eds: P L DE BOER, A VAN GELDER AND S D NIO). Reidel, Dordrecht (in press).

FLEMMING, B W and A B THUM (1978). The settling tube - a hydraulic method for grain size analysis of sands. **Kieler Meeresforschungen, Sonderheft,** 4, 82-95.

FLEMMING, B W and E R HAY (1983). On the bulk density of South African marine sands. **Joint Geological Survey/University of Cape Town Marine Geoscience Unit, Technical Report** 14, 171-176.

FLEMMING, B W and E R HAY (1984). On the bulk density of South African marine sands. **Transactions of the Geological Society of South Africa,** 87, 233-236.

FLEMMING, B W and A K MARTIN (1985). Nearshore submerged spit-bars: a facies model. **Terra cognita,** 5, p.60.

GIBBS, R J (1981). Sites of river-derived sedimentation in the ocean. Geology, 9, 77-80.

GILL, A E A and E H SCHUMANN (1979). topographically induced changes in the structure of an inertial coastal jet: application to the Agulhas Current. **Journal of Physical Oceanography,** 9, 975-991.

GOODLAD, S W (1986). Tectonic and sedimentary history of the mid-Natal Valley (SW Indian Ocean). PhD thesis, (unpublished), Department of Geology, University of Cape Town, South Africa. 415 pp.

GRÜNDLINGH, M L (1986). Features of the northern Agulhas Current in Spring 1983. **South African Journal of Science,** 82, 18-20.

HARMS, J C, J B SOUTHARD and R G WALKER (1982). Structures and sequences in clastic rocks. **SEPM Short Course, No. 9,** 249 pp.

HAY, E R (1984). Sediment dynamics on the continental shelf between Durban and Port St Johns (Southeast African Continental Margin). MSc thesis, (unpublished), Department of Geology, University of Cape Town, South Africa. 238 pp.

MARTIN, A K (1984). Plate tectonic status and sedimentary basin in-fill of the Natal Valley (SW Indian Ocean). **Joint Geological Survey/University of Cape Town, South Africa. Marine Geoscience Unit, Bulletin No. 14,** 209 pp.

MARTIN, A K (1985). The distribution and thickness of Holocene sediments on the Zululand continental shelf between 28° and 29°S. CSIR, South Africa, Report C/SEA 8524, 13 pp.

McCAVE, I N (1985). Recent shelf clastic sediments. **Geol. Soc. London, Spec. Publ.** 18, 49-65.

MIDGLEY, D C and M V PITMAN (1969). Surface water resources of South Africa. **Hydrological Research Unit, Department Civil Engineering, Witwatersrand, South Africa. Report No. 2/69** 128 pp.

MOIR, G J (1976). Preliminary tectural and compositional analyses of surficial sediments from the upper continental margin between Cape Recife (34°S) and Ponta do Ouro (27°S), South Africa. **Joint Geological Survey/University of Cape Town, South Africa. Marine Geology Programme, Technical Report No. 8.** 68-75.

MURGATROYD, A L (1979). Geologically normal and accelerated rates of erosion in Natal. **South African Journal of Science,** 75, 395-396.

NICHOLSON, J (1983). Sedimentary aspects of the Mvumase Project. In: **Beaches as ecosystems.** (Eds: A McLACHLAN and T ERASMUS). Junk, The Hague, 191-197.

ORME, A R (1973). Barrier and lagoon systems along the Zululand coast. In: **Coastal geomorphology.** (Ed) D R COATES, State University, New York, Binghampton, 181-217.

PEARCE, A F, E H SCHUMANN and G S A J LUNDIE (1978). Features of the shelf circulation off the Natal coast. **South African Journal of Science,** 74, 328-331.

ROOSEBOOM, A (1978). Sedimentafvoer in Suider-Afrikaanse riviere. **Water South Africa,** 4, 14-17.

ROOSEBOOM, A (1982). Interim report on expected sediment-related changes due to the Mvumase Scheme. **Department of Environment Affairs,** unpublished.

SCHWARTZ, H I and R A PULLEN (1966). A guide to the estimation of sediment yield in South Africa. **Transactions of the South African Institute of Civil Engineering,** 8.

SIESSER, W G, R A SCRUTTON and E S W SIMPSON (1974). Atlantic and Indian ocean margins of Southern Africa. In: **The Geology of Continental Margins.** (Eds: C A BURK and E L DRAKE). Springer Berlin, 641-654.

WALKER, R G (1979. Shallow marine sands. In: **Facies Models.** (Ed) R G WALKER, Geoscience Canada, Reprint Series 1, 75-90.

WEISSER, P J, I F GARLAND and B K DREWS (1982). Dune advancement 1937-1977 at the Mlalazi Nature Reserve Mtunzini, Natal, South Africa. **Bothalia** 14, 127-130.

WEISSER, P J and A P BAKER (1983). Monitoring beach and dune advancement and vegetation changes 1973-1977 at the farm Twinstreams, Mzunzini, Natal, South Africa. In: **Sandy Beaches as Ecosystems.** (Eds: A McLACHLAN and T ERASMUS). Junk, The Hague, 727-732.

Chapter 4

CLIMATE AND WEATHER OFF NATAL

Ian T Hunter
National Research Institute for Oceanology
Council for Scientific and Industrial Research

THE CLIMATE OF NATAL VIEWED ON THE GLOBAL SCALE

According to the Köppen Climate Classification (Boucher, 1975), the climate of the Natal coastal belt is given the symbol Ca, that is, it is a humid sub-tropical climate with a warm summer. In the Southern Hemisphere it shares this classification with a portion of the east coast of Australia and the north coast of Argentina. These three areas are all situated on eastern seaboards and centred at about 30°S. This latitude is an important common factor since the southern sub-tropical high-pressure belt (STHP) has its mean position on the 30th parallel and therefore plays a dominant role in this climatic zone.

Mean sea-level pressure analyses of the Southern Hemisphere generally show the STHP belt to comprise several separate oceanic cells. Taljaard (1972) found the highest concentration of anticyclones to be centred at approximately 90°W, 10°W and 100°E, and these may be regarded as the mean positions of the three semi-permanent STHP systems of the Southern Hemisphere. They are warm-cored and thus dominate the circulation throughout the troposphere, tilting towards the equator with height.

All three 'Ca' regions lie to the east of an STHP system, with various amounts of land mass and ocean separating them from the semi-permanent cells. It is of interest to note that conditions are very arid on the western coasts of these continents where the close proximity of a major anticyclone results in strong subsidence throughout the year.

Yet it is these same STHP systems which regularly 'bud off' migratory highs to dominate east coast climates. This budding-off process plays a vital role in the weather processes of all three eastern seaboards.

The 'bud-off high' (BOH) originates when the STHP cell extends a ridge eastward; in the African case this is usually just south of the sub-continent in summer but often overland in winter when the main cell is approximately 3° to 4° further north. In South America ridging takes place across the Andes in all seasons (Taljaard, 1972). When the BOH forms, the parent cell retreats towards its former position, and the new cell continues eastward. What appears to be another semi-permanent STHP cell east of Durban, the so-called Indian Ocean High (IOH), is simply a composite of eastward-moving migratory highs.

THE CLIMATE OF THE NATAL COAST

Sources of Climate and Weather Data

In order to be able to portray adequately the climate of any region, a source of long-term meteorological data is required. Along the Natal coast, Louis Botha airport about 15 km south of Durban is the only station at which long-term surface measurements of climatic conditions are being made hourly, although until 1957 measurements had been made at the Stamford Hill aerodrome closer to the city centre. Louis Botha airport is also the only source of long-term radiosonde measurements on the Natal coast. The only other long-term surface observations are contained in synoptic reports from the lighthouse keepers at Port Shepstone and Cape St Lucia (Fig. 1.1).

Data from voluntary observing ships (VOS) reports were also used, with 42 479 observations available for the area 27°S to 31°S and 30°E to 34°E (1962-1979). These were provided by the South African Data Centre for Oceanography (SADCO).

Mean Wind Circulation

Figures 4.1 and 4.2 depict average wind roses for the three sites mentioned above. The coastline lies roughly northeast/southwest, and the dominance of coastwise winds is apparent, especially in summer. Over the southern parts of the Natal coast, southwesterly and northeasterly winds are roughly balanced in frequency, but in the

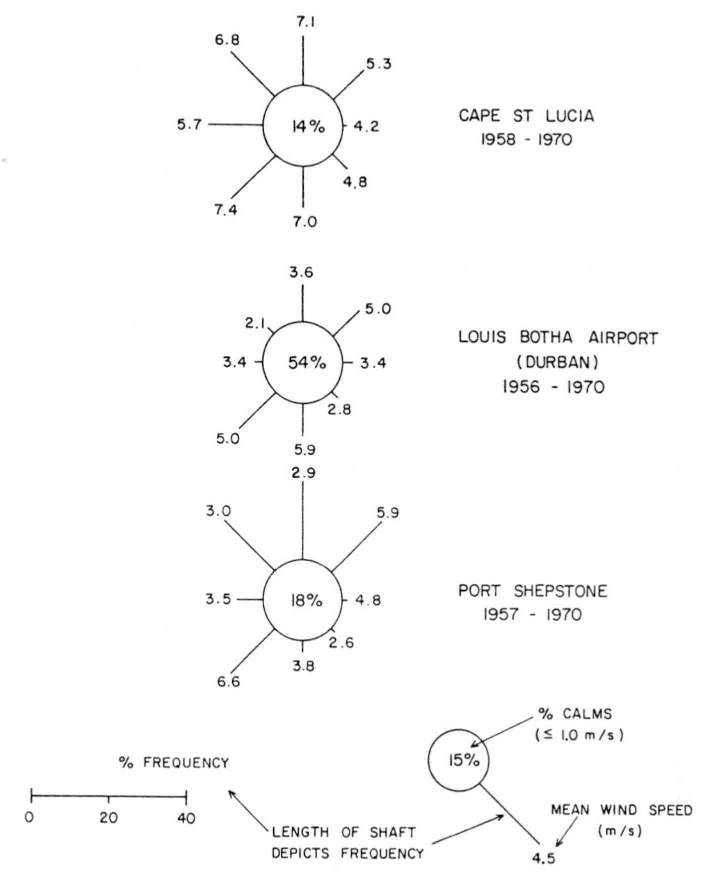

Figure 4.1 Average wind conditions along the Natal coast in June. The lengths of the records in each case are given.

north, northeasterly winds dominate the summer wind rose. The true directions of these prevailing winds are NNE and SSW. The higher percentage of calms at Louis Botha Airport is the direct result of its location, as it is separated from the coast by a 90 m high bluff.

The average wind roses for June show a significant increase in offshore flow at Cape St Lucia and Port Shepstone. The effect of the poor location of the Louis Botha anemometer is again evident with almost no offshore flow even though data from other anemometer sites in Durban area prove the existence of a marked offshore flow in autumn and winter (Lundie, 1979). Although the majority of offshore winds in winter are related to the nocturnal land breeze, northwesters may also

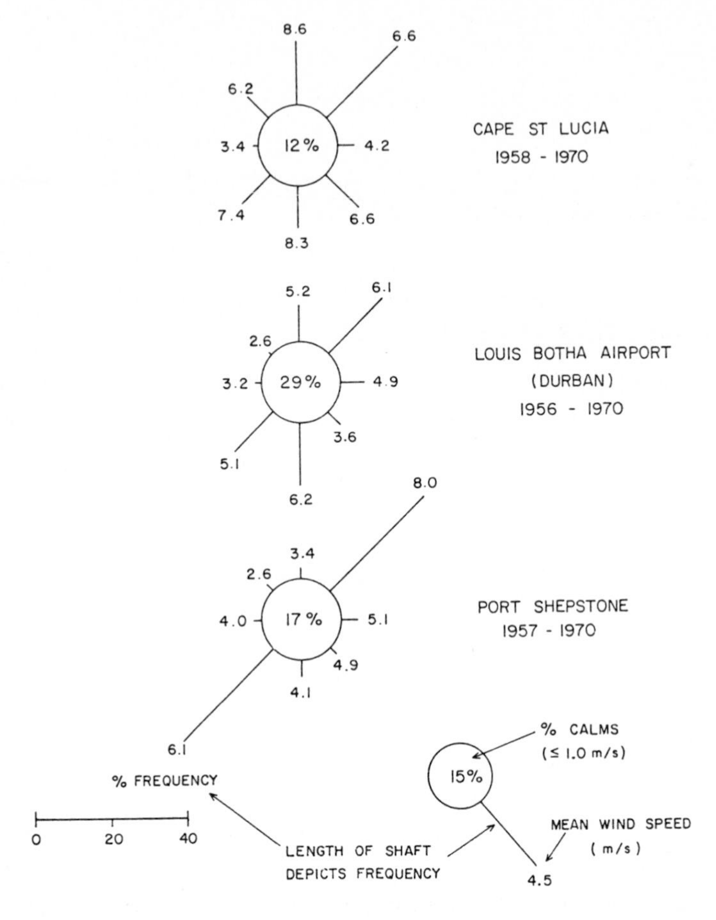

Figure 4.2 Average wind conditions along the Natal coast in December.

occur under bergwind conditions (see next Section).

A search of the VOS data set showed that southwesterly through to southerly winds generally have the highest speeds. As may be expected, wind conditions vary greatly offshore. The highest hourly (average) wind speed to be measured at Louis Botha Airport in the period 1956 to 1970 was 24.6 m/s. This was observed in November, but on average it is September which is the stormiest month.

June is the month with the least wind, both in terms of lowest frequency of gales and highest percentage calms (South African Weather Bureau, 1975).

Air Temperature

The air temperature field is characterized by a relatively low seasonal range (see Table 4.1) which is due to the damping effect of the adjacent ocean. On the diurnal scale, however, large variations may occur, especially when bergwind conditions occur. The absolute maximum temperature of 42°C, given in Table 4.1, was caused by such a condition. The absolute minimum temperature of 4°C refers to a screen height of 1.3 m. Light frost has however been reported at Louis Botha Airport where the absolute minimum screen temperature is 2.8°C (South African Weather Bureau, 1963).

The VOS data set shows absolute minimum air temperatures offshore tending towards about 10°C, and maximum values offshore between 33°C and 35°C. In the latter case, all sea surface temperature (SST) measurements exceeded 25°C. This is to be expected since lower SST values would have a significantly moderating effect on the warm subsiding air.

Relative Humidity

The annual average relative humidity of 79 per cent does not endear Durban to many South Africans. Because of the close proximity of the warm Agulhas current, the highest moisture levels in the country occur along the Natal coastal belt. The sea breeze is responsible for carrying large amounts of moisture to the coast, and relative humidity usually rises rapidly after 09h00. On the other hand, the land breeze brings in dry air from the interior and under bergwind conditions relative humidities may drop to below 30 per cent (South African Weather Bureau, 1974).

Sea Level Pressure

Table 4.1 shows that monthly mean atmospheric pressures are seen to be significantly higher in winter than in summer. This is due to an increase in the average intensity of the migratory highs which also track closer to the Natal coast during the winter months. VOS reports give an absolute minimum pressure of 992 mb offshore, with an absolute

Table 4.1 Climatic data from Durban's old municipal aerodrome at Stamford Hill. This site lies at an elevation of 5 m and is strongly influenced by marine conditions. (from Schulze, 1984).

Month	Mean sea level press. (mbar)	Temperature (°C) mean max.	mean min.	extreme max.	extreme min.	Rel. humid (%)	Precipitation mean (mm)	max. (mm)	min. (mm)	days 1 (mm)	max. in 24 h (mm)
Jan.	1 014	27	20	33	14	81	118	383	10	11	177
Feb.	1 014	28	21	38	15	82	128	358	22	9	151
Mar.	1 015	27	20	32	14	83	113	267	23	9	83
Apr.	1 016	26	18	37	11	81	91	315	8	7	146
May	1 018	24	14	35	7	77	59	260	5	4	161
June	1 022	22	11	32	5	73	36	356	0	3	240
July	1 022	22	11	33	4	73	26	109	1	3	55
Aug.	1 020	22	12	36	5	77	39	136	2	4	82
Sept.	1 020	23	15	42	8	79	63	143	7	6	95
Oct.	1 017	23	17	40	8	81	85	251	25	10	78
Nov.	1 015	25	18	39	10	81	121	278	21	11	135
Dec.	1 014	26	19	35	13	81	124	363	41	12	100
Annual	1 017	25	16	42	4	79	1 003	1 397	631	89	240
Record (yrs.)	30	30	30	20	20	20	30	30	30	30	30

Month	Average Wind Speed (km/h)	Averages: Cloudiness (tenths) 08h	Cloudiness (tenths) 14h	Sunshine (h/day)	Radiation (cal/cm² / day)	Diffuse Radiation	Evaporation (mm)
Jan.	12	6	6	6.5	520	206	203
Feb.	11	5	5	6.7	480	188	170
Mar.	11	5	5	6.2	434	152	180
Apr.	9	4	4	6.8	365	112	127
May	8	3	4	7.4	266	83	99
June	8	3	3	6.9	275	70	84
July	9	3	3	6.9	285	76	89
Aug.	12	4	3	6.9	324	103	122
Sept.	13	5	5	5.8	402	144	142
Oct.	14	6	7	5.2	413	178	170
Nov.	15	6	7	5.7	459	215	173
Dec.	13	6	6	6.0	495	233	196
Annual	11	5	5	6.4	393	147	1 755
Record (yrs)	10	30	30	10	7	7	4

maximum near 1 040 mb; these seaward observations are likely to have been made closer to the centres of the high- or low-pressure systems. However, VOS reports are by no means continuous, so that more extreme pressures may well have occurred when no voluntary observing ships were in the vicinity.

Cloud and Solar Radiation

Average cloudiness at Louis Botha Airport (South African Weather Bureau, 1974) reaches a maximum in summer (5.5 oktas), with winter showing the least cloud (2.6 oktas). Its annual average of 4.4 oktas makes Durban, which can be taken as representative of the whole Natal coast, one of the cloudiest places on the sub-continent. This is reflected in the figures for average annual total solar radiation which reaches a minimum on the Natal coast, Durban receiving 393 cal cm^{-2} day^{-1}, while that portion due to diffuse radiation reaches a maximum of 147 cal cm^{-2} day^{-1}, (Table 4.1).

The effect of the Agulhas current on the development of cumulus congestus cloud offshore is frequently visible on satellite images (e.g. see Figure 5.14); narrow cloud lines often stretch for hundreds of kilometres in its vicinity. Studies have shown that it is the migratory high which is most often linked to such cloud formation (Lutjeharms _et al_., 1985). As will be seen in the next section, these cloud banks sometimes affect coastal weather.

Precipitation

Mean annual precipitation varies between 1 000 mm and 1 100 mm over much of the coast (South African Weather Bureau, 1965), the exception being the extreme northern parts where Kosi Bay receives 980 mm, and the Cape St Lucia area where the rainfall is over 1 200 mm in places.

The rainfall anomaly in the Richards Bay/Cape St Lucia portion of the coastal belt has resulted in one climate classification by Linton (1975) which separates this area from the rest of the coastal belt. However, it is highly likely that the increased total rainfall is due to the close proximity of the Agulhas current with its attendant cumulus cloud. As might be expected, the anomaly is largely the result of relatively high winter rainfall when heavy post-frontal showers are often unique to this area.

Over the rest of the coastal belt winter monthly rainfall figures are typically less than 30 per cent of the summer values. For the whole coastal belt, precipitation reaches its maximum in the summer months, November to March.

SYNOPTIC WEATHER SYSTEMS - THE BASIC CYCLE

Although a great variety of weather combinations may affect the Natal coast at any one time, a basic weather cycle is clearly discernible in the long-term. This is depicted in Figure 4.3 which shows the movement and development of synoptic scale systems through one cycle, as well as a multiple time series over the same period as recorded by an automatic weather station on Durban's Bluff. The time series of wind speed includes an average over the measurement cycle (15 minutes), as well as gust, which is the maximum 2-second speed in this period.

Periodicity

Using five years of hourly atmospheric pressure values from Louis Botha Airport, Hunter (1984) produced surface pressure spectra with peaks at 3.8 and 5.9 days. A 3-hourly filtered time series of the same data set (Figure 4.4), and reference to the relevant synoptic charts, indicates that the 3.8 day period was of marked synoptic significance. Previously only the 5.9 day period had been recognised (Preston-Whyte and Tyson, 1973). Figure 4.4 shows that weather systems may pass along the Natal coast with great regularity for up to 2 weeks. Periodicities of 3 days or less are quite common, while it is unusual for a weather cycle to take longer than a week to complete.

Example of a Basic Weather Cycle

The so-called "coastal low" has probably received more attention than any other weather system affecting this coastline. Its formation is due to the interaction between large-scale atmospheric flow and the marked Southern African escarpment (De Wet, 1984). It is a shallow system capped by a subsidence inversion at about 1 km. Another characteristic is a vertical wind shear near the inversion (Estie, 1984). These systems propagate around Southern Africa in an

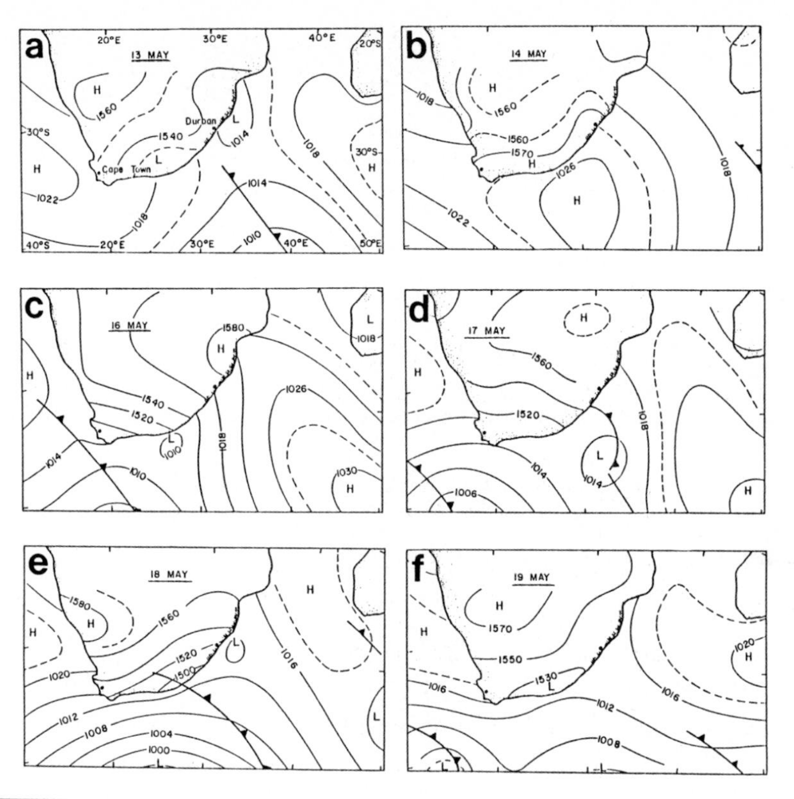

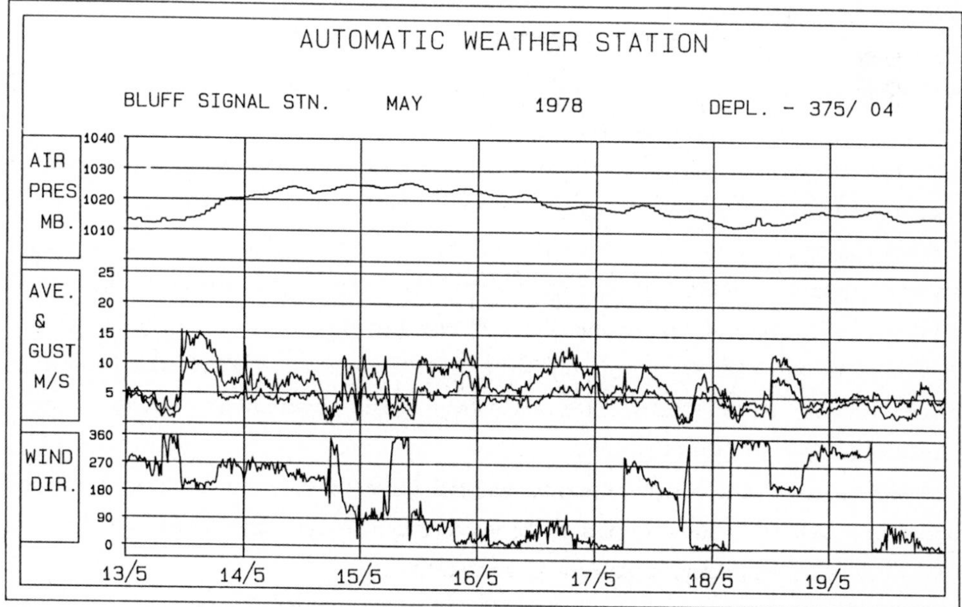

Figure 4.3 A typical weather cycle during 1978, as depicted by 12h00 G.M.T. charts (upper figures), and multiple time series from Louis Botha Airport (hourly air pressure values) and a station on the Bluff (15-minute wind recordings).

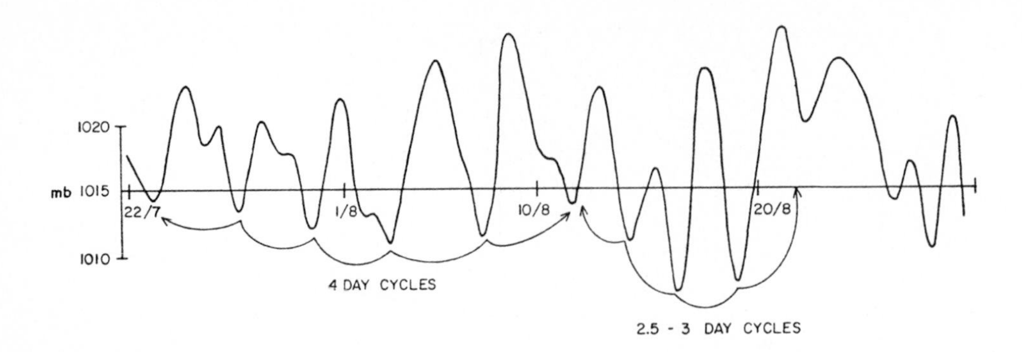

Figure 4.4 Filtered hourly air pressures from Louis Botha Airport, Durban over the period 22 July to 30 August, 1978.

anti-clockwise direction, and have been modelled by Gill (1977) as coastally trapped waves in the atmosphere.

The time series in Figure 4.3 represent a basic weather cycle, which are now discussed in conjunction with the relevant synoptic charts.

13 May 1978. At 11h00 a coastal low passed through Durban. Its passage was well-defined by a 'buster' or gust front, associated with very rapid changes of wind direction and speed. This phenomenon has, in the past, been recorded 50 km offshore in association with such coastal lows (Marine Observer, October 1974). Note that in summer the pre-coastal low winds can be of gale force NNE, especially offshore. In this autumn example, a land breeze is able to dominate the pre-coastal low wind field.

14 May 1978. The important budding-off process mentioned earlier resulted in a separate high-pressure cell moving up the coast. (At times, particularly in the winter, this bud-off high may simply appear off the Natal coast after the Atlantic High has ridged overland). Winds eventually backed round to ESE off Natal.

15 and 16 May 1978. The bud-off high now merged with the so-called Indian Ocean High (IOH) and easterly wind components were largely modulated by the land breeze.

17 May 1978. The next coastal low to move up the coast went offshore south of Durban, and formed a wave on the associated front. Typical post-coastal low winds did not materialize.

18 May 1978. The sudden onset of SSW winds at 13h00 indicated the passage of a coastal low through Durban, more than 5 days behind its predecessor. Note that major changes in pressure, temperature and wind

on the Natal coast do not occur with the passage of a cold front, but with a coastal low moving through.

19 May 1978. There was no longer any sign of the coastal low and the associated front was well to the south-east.

Each of the three coastal lows depicted in Figure 4.3 had a 'travelling companion' in the form of a cold front, usually well to the south and moving in a southeasterly direction. One variation of the above pattern occurs when the cold front moves in a northeasterly direction to sweep over much of the sub-continent. In an extreme case in June 1964, a cut-off low developed behind such a front, and thick snow fell within 100 km of the coast.

As shown on 19 May, not every coastal low/front combination is followed by a ridge of the Atlantic High. The situation in which a coastal low moves offshore is actually more common to the north of Durban. These systems may also simply fill up before reaching the Natal coast.

Associated Meteorological Conditions

Thunderstorms. Thunder is recorded on an average of about 30 days per annum on the Natal coast. Most thunderstorms reach the coast from the interior when the pre-coastal low air is unstable. However, at times (Jackson, 1964) cumulo-nimbus cells form offshore over the Agulhas current, and move over the coast in the early hours of the morning. It is possible that the land breeze plays a role in triggering this phenomenon. Vigorous thunderstorms have been observed 100 km off Richards Bay under synoptic conditions which favoured land breeze development (Marine Observer, April 1978).

Bergwind conditions. This name is misleading since the term refers more to a temperature condition than to a specific wind. The main requirement is subsiding air with adiabatic heating producing abnormally high surface temperatures. This usually occurs when a coastal low approaches the Natal coast, and the subsidence inversion extends down to the surface. This situation arises with a relatively small percentage of coastal lows, however, with the Natal coast receiving less than 30 days per annum with maximum temperatures above 30°C.

Orographic rain. A large percentage of the rainfall on the Natal coastal belt occurs when the Atlantic High extends a ridge up the coast behind a coastal low/front combination (Figure 4.3b). Eventually a separate cell of high pressure passes to the south, and onshore flow ceases, causing a clearance over the coast. Since it is a very shallow system, the coastal low is not directly associated with any significant precipitation.

Fog. This is a rare condition on the Natal coast, being recorded on less than 2 days per annum at Durban's Point (Great Britain Admiralty, 1944). A search of the VOS reports produced only six fog observations (visibility less than 1 km), all in Spring or Summer. On all six occasions sea surface temperatures were well above 20°C. Since suitably cold water is seldom available along this coast, advection fog is unlikely. A possible fog-creating mechanism is that dry bergwinds blowing over the warm waters of the Agulhas current pick up considerable amounts of moisture, after which mixing with cooler air could then bring this air to its dew point. In five out of the above six instances bergwind conditions were experienced on the coast prior to the observation of fog. A major factor in the maintenance of fog is the presence of an inversion inhibiting mixing in the vertical.

RECURRING SYNOPTIC PATTERNS OF MARKED METEOROLOGICAL AND OCEANOGRAPHIC SIGNIFICANCE

The basic weather pattern described in the previous Section is sometimes masked by a more dominant synoptic situation. The first three synoptic conditions depicted in Figure 4.5 are associated with significant weather events which recur every year; the last is a much less common occurrence.

Cyclogenesis and an Intense Migratory High

Figure 4.5a. This is simply a further development of part d of the basic cycle (Figure 4.3d). The deep wave located south-east of Durban undergoes rapid cyclogenesis as it moves further east. With a bud-off high ridging in behind the frontal trough, coastwise pressure gradients

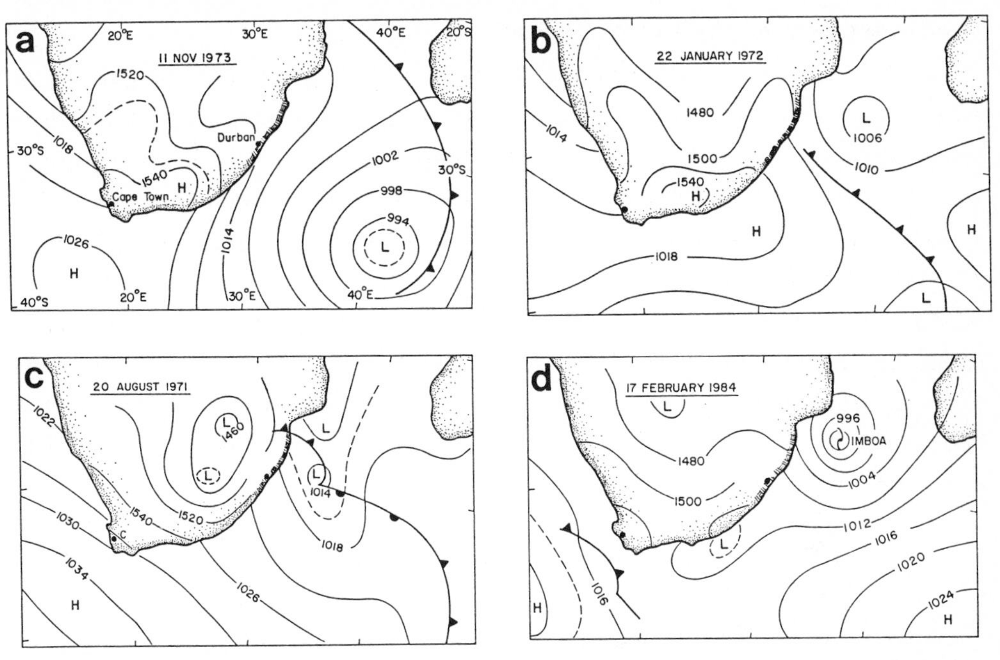

Figure 4.5 Synoptic patterns related to extreme meteorological and oceanographic conditions on the Natal coast. Synoptic hour, 12h00 G.M.T.

are tight, and gale force winds occur.

This is a combination which is well-known for its generation of high winds and heavy swell. On 11 November a vessel 50 km east of Durban reported SSW winds of 25 m/s and a southwesterly swell of 13.5 m. Similar synoptic conditions resulted in the sinking of the Mauritius II in April 1980 and the Tong Nam in July 1979, both in an area roughly 100 km north-east of Durban (South African Shipping News and Fishing Industry Review, September 1979 and May 1980).

On 27 June 1981, with synoptic conditions basically similar to those above, the MV Pacific Crane reported an encounter with a "freak wave" (Marine Observer, April 1982). The vessel was in the Agulhas Current core (estimated at 8 knots), about 1 nautical mile offshore of the 200 m isobath, when a 'wall of white water' covered the forecastle. Although the general swell field was reported to be southwesterly 8 to 10 m, the rogue wave was estimated to be 15 to 18 m high. Such waves occur as the result of southwesterly swell interacting with the Agulhas Current (Schumann, 1980).

Stationary Low North-East of Durban

Figure 4.5b. It is not uncommon for a low to appear north-east of Durban at the time that the coastal low disappears from the coast. It would therefore seem likely that the offshore system has its origins in the coastal low, although its dynamics must differ significantly.

This occurred on 22 and 23 January 1972. Such systems may remain stationary for several days, causing a tight coastwise gradient to build up as the Atlantic high ridges in; they may also intensify markedly in this position. On 22 January a maximum southerly swell of 6 m was reported by a vessel in the vicinity of Port Shepstone. A southerly wind of 27.5 m/s was reported from just south of Kosi Bay early the following day.

Cut-off Low

Figure 4.5c. The phenomenon shown here over the eastern interior is known as a 'cut-off low'. The term refers to a condition in which cold air which was originally part of a mid-latitude trough, is cut-off by a ridge of relatively warm air. The condition is usually reflected at the surface with the Atlantic High extending a deep warm ridge well to the south.

The low, being cold-cored, is a deep system which is well-defined through the troposphere. It is associated with a strong circulation which has been responsible for many floods in South Africa, as occurred on 20 and 21 August, 1971, resulting in a 24 hour rainfall figure of 235 mm at Port Shepstone.

According to Taljaard (1982), the number of cut-off lows affecting the sub-continent in a year ranges from 5 to 15. The majority move offshore south of Natal, but there are cases where cut-off lows have passed north of Durban. De Villiers (1977) gives two examples of flood damage in the Durban area, both associated with cut-off low conditions. Cut-off lows do not necessarily form overland. On 18 May 1974 a cut-off low developed approximately 200 km seaward of Cape St Lucia causing SSW winds of 30 m/s, and heavy rain inshore (VOS report). Swell was estimated to be 9 m (SSW). A cut-off low may cause a strong onshore flow, depending on its relative position (known as a 'black south-easter' because of its contrast with the transient south-easter of fair weather).

Referring again to Figure 4.5c, the strong ridge which separates the cut-off low from its original westerly flow may form a temporary obstruction to weather systems moving eastwards; this is known as a 'blocking' high.

Apart from possible floods, gale-force winds and heavy swell, cut-off lows may be associated with other phenomena offshore. On 26 December 1974 a cut-off low was moving seaward south of Durban. A vessel 40 km SSW of Richards Bay experienced a vigorous line squall with a SSW buster reaching Force 9 (Marine Observer, October 1975). Thunder occurred and a group of shallow water spouts were observed. This system was not particularly active over land.

Tropical Cyclones

Figure 4.5d. Although the sight of a tropical cyclone threatening northern Natal is not a common one, this pattern assumed particular significance in January and February 1984 when two tropical systems struck within a fortnight, causing extensive damage. Prior to this, only one case of severe flooding on the Natal coast during the previous 20 years could be ascribed to a tropical cyclone, although wave damage to coastal structures has occurred more frequently.

Tropical cyclone 'Imboa' caused the pressure to drop to 990 mb at Cape St Lucia on 18 February (Poolman and Terblanche, 1984); this pressure is actually lower than the absolute minimum contained in the VOS data set. It was responsible for rainfalls of more than 300 mm in the Richards Bay area. The Richards Bay harbour was closed on 18 February due to a 10 m swell, while wave damage was extensive as far south as Durban. Gale force southwesterly to southeasterly winds were experienced.

Prior to Imboa, tropical cyclone Domoina had already caused much devastation over northern Natal, where several bridges were swept away. Rainfall on the coast in the Cape St Lucia/Richards Bay area exceeded 400 mm.

The summer of 1984 was by no means average as far as tropical cyclones are concerned. Ten systems formed in the south-western Indian

Ocean compared with a seasonal average of six. Most tropical cyclones develop north-east of Madagascar between November and April, and although an average of one may occur each year on the southern Mozambique coast, the number for Natal is much less than this.

LAND AND SEA BREEZES

Most of the possible synoptic-scale situations have been discussed. Superimposed upon these circulations, and sometimes swamped by them, are the locally generated land and sea breezes. Though they are relatively weak, they occur frequently and cannot be ignored. Sonu *et al*. (1973) have emphasized the folly of disregarding the effect of this 'diurnal coastal air circulation' on the coastal ocean.

The Sea Breeze

Unlike the west coast of South Africa where land/sea temperature contrasts may exceed 20°C in summer, the Natal coast, having a much warmer body of seawater washing its shores, does not have a very strong sea breeze circulation. Prevailing gradient winds usually have an onshore component prior to the onset of the sea breeze, which prevents the formation of steep temperature gradients. As a result, the development of a sea breeze front with rapid temperature and wind changes is a rare phenomenon on the Natal coast (Preston-Whyte, 1969).

The sea breeze begins in the summer at about 09h00 and continues until approximately 20h00 (Jackson, 1954). In winter, when the land/sea temperature contrast is favourable for a much shorter time, it blows from about 11h00 to 17h00 and is much weaker. Because of frictional effects (Preston-Whyte, 1969) maximum speed is usually attained some time after the maximum air temperature is reached. Land and sea breezes are seldom devoid of any gradient wind influence, and the sea breeze is most prominent when the large-scale circulation is north-east. This is due to generally greater land/sea temperature gradients under these fine weather conditions. Southwesterly winds, on the other hand, are often associated with cloud and tight pressure gradients, and show little change in direction during the day

(Smith, 1961).

Land Breeze

This circulation has been greatly underestimated in the past. This was because, first, Durban's main reference meteorological station Louis Botha Airport, receives virtually no offshore flow due to the surrounding topography and second, because offshore conditions had not been studied.

The wind roses in Figure 4.1 (June), testify to the importance of offshore flow along the Natal coast in winter; Lundie (1979) showed the true situation with readings from an automatic weather station on Durban's Bluff. From these observations it was clear that the northwesterly land breeze represents a significant amount of the total kinetic energy in winter, comparable with that of the coastwise gradient winds.

Hunter (1981) used offshore wind observations to show that the land breeze extends to at least 60 km off the Natal coast in winter. This was ascribed to a strong temperature gradient between the coast and the Agulhas current. Speeds of over 9 m/s were recorded, 20 km from the coast.

Using Bluff anemometer data, Hunter (1981) found an average starting time of 19h00. The land breeze usually dissipates by 08h00, with the offshore component peaking at about 07h00, that is, approximately half an hour after sunrise. Preston-Whyte (1974) emphasized the importance of a mountain-plain wind linked up with the coastal land-breeze circulation during the night.

Examples of Land and Sea Breezes

Several occurrences of land and sea breezes are depicted in Figure 4.3. On 16 May the NNE gradient wind veered to ENE as the sea breeze increased in strength. Maximum deviation from the coast occurred at 16h30 but the maximum speed of 6.9 m/s occurred at 17h30, presumably more a function of the large-scale pressure field, than of the local influence. Preston-Whyte (1969) found the sea breeze in the Durban area to be relatively shallow (less than 1 km deep), with a maximum flow below 300 m and liable to surging.

On 18 May 1978 at approximately 19h00, a SSW wind veered to a northwesterly land breeze which continued through the night, reaching a peak of 4.8 m/s at 07h00. The subsequent clockwise rotation of the wind vector was due to a combination of weakening land breeze, NNE gradient wind and intensifying sea breeze.

CONCLUSION

The general description of weather conditions along the Natal coast has covered atmospheric phenomena occurring on all time scales from climatic through synoptic to diurnal. However, particular emphasis has been placed on the synoptic scale weather systems, in contrast to the usual approach to such a study wherein the emphasis is on the climatic time scale.

In terms of both its weather and its climate this region is unique in the Southern Hemisphere, due mainly to a combination of an intense western boundary current and a hinterland rising to over 3 000 m above sea level. The coastal low for example, is partially a product of the high interior escarpment, and has been shown to play a major role in the region's weather. At present very little is known about its offshore character and particularly the associated buster.

The Agulhas Current, on the other hand, gives the region an enhanced land breeze circulation offshore; there may also be a possible link between this circulation and the initiation of convection both over the Agulhas Current and, at times, over the coast.

Due to its latitudinal position, the Natal coastal belt comes under the influence of both temperate and tropical weather systems. Intense frontal systems coupled with the poleward-flowing Agulhas Current can cause the high energy swell that is renown amongst the world's mariners. Tropical cyclones are less frequent than on (for instance) the Australian east coast, but a single event may cause coastal damage which remains apparent for many years afterwards.

This whole interaction between atmospheric and oceaning events is an issue which requires more clarification, particularly the possibility that this is a source of energy not only for mesoscale systems, but is also a major factor in the intensification of synoptic scale low pressure systems.

REFERENCES

BOUCHER, K (1975). **Global Climate.** The English Universities Press Ltd., London.

DE VILLIERS, M P (1977). Localised flooding in Durban due to thunderstorms. **South African Weather Bureau Newsletter** 337, 117-120.

DE VILLIERS, M P (1978). The Durban Storms - 29 and 30 December 1977. **South African Weather Bureau Newsletter** 347, 33-38.

DE WET, L W (1984). The dynamic forcing of coastal lows. Coastal Low Workshop, Simonstown. Unpublished.

ESTIE, K E (1984). Forecasting the formation and movement of the coastal low. Coastal Low Workshop, Simonstown. Unpublished.

GILL, A E (1977). Coastally-trapped waves in the atmosphere. **Quarterly Journal of the Royal Meteorological Society,** 103, 431-440.

GREAT BRITAIN, ADMIRALTY (1944). Meteorological services of the Royal Navy and Union of South Africa. **Weather on the coasts of Southern Africa,** Vol. 2.

HUNTER, I T (1981). On the land breeze circulation of the Natal coast. **South African Journal of Science,** 77, 376-378.

HUNTER, I T (1984). Coastal lows from a synoptic point of view. Coastal Low Workshop, Simonstown. Unpublished.

JACKSON, S P (1954). Sea breezes in South Africa. **South African Geographical Journal,** 36, 13-23.

JACKSON, J K (1964). Severe thunderstorm at Durban. **South African Weather Bureau Newsletter,** 181, 85-86.

LINTON, D L (1975). **Climate Classification.** New Oxford Atlas. Oxford University Press.

LUNDIE, G S H (1979). Land breeze contributions to the wind field in the Durban area. NRIO (CSIR) Memo 7913.

LUTJEHARMS, J R E, R D MEY and I T HUNTER (1985). Cloud lines over the Agulhas current. Unpublished manuscript.

PRESTON-WHYTE, R A (1969). Sea breeze studies in Natal. **South African**

Geographical Journal, 51, 38-49.

PRESTON-WHYTE, R A and P D TYSON (1973). Note on pressure oscillations over South Africa. **Monthly Weather Review,** 101, 650-659.

PRESTON-WHYTE, R A (1974). Land breezes and mountain-plain winds over the Natal coast. **South African Geographical Journal,** 56(1), 27-35.

SCHULZE, B R (1984). World survey of climatology. Volume 10, **Climates of Africa,** Chapter 15. Elsevier, Amsterdam.

SCHUMANN, E H (1980). Giant wave: anomalous seas of the Agulhas Current. **Oceans,** 13, 27-30.

SMITH, A J J (1961). Marine studies off the Natal coast. CSIR Symposium, Pretoria, South Africa.

SONU, C J, S P MURRY, S A HSU, J N SUHAYDA and E WADDELL (1973). Sea breeze and coastal processes. **EOS** Vol. 54(9), 820-833.

SOUTH AFRICAN WEATHER BUREAU (1963). Aeronautical climatological summary: Louis Botha Airport - Durban. **Newsletter,** 177, 213.

SOUTH AFRICAN WEATHER BUREAU (1974). **Climate of South Africa, Part 8,** Pretoria.

SOUTH AFRICAN WEATHER BUREAU (1965). **Climate of South Africa, Part 9,** Pretoria.

SOUTH AFRICAN WEATHER BUREAU (1975). **Climate of South Africa, Part 12,** Pretoria.

TALJAARD, J J (1972). Synoptic meteorology of the southern hemisphere. **Meteorological Monographs,** Vol. 13(15). American Meteorological Society, Boston, Mass.

TALJAARD, J J (1982). Afknyplae oor Suid-Afrika en omgewing. **South African Weather Bureau Newsletter,** 396, 33-37.

TILBURY, M (1980). Sikloon Kolia veroorsaak buitengewone deining aan die Natalse Kus. **South African Weather Bureau Newsletter,** 374, 80-82.

Chapter 5

PHYSICAL OCEANOGRAPHY OFF NATAL

Eckart H Schumann
Department of Oceanography
University of Port Elizabeth

GENERAL DESCRIPTION

A description of the origins and structure of the continental shelf off Natal has been given in Chapter 2. The nature of this shelf has a major influence on the ocean dynamics, delineating regions with markedly different characteristics.

Further offshore, the most important large-scale oceanographic feature is undoubtedly the Agulhas Current. It is generally accepted that this classical western boundary current forms off the Northern Natal/Mozambique coast, from a confluence of waters which follow complex paths in the Mozambique Channel and areas south of Madagascar (Lutjeharms, Bang and Duncan, 1981, Saetre and Da Silva, 1984, Gründlingh and Pearce, 1984). It is recognised as one of the world's major currents, and sweeps polewards with the core generally just offshore of the shelf break. As such, the waters on the shelf are affected markedly, and an understanding of this current is necessary before the processes operating on the shelf can be fully understood.

Weather, and particularly the wind, play fundamental roles in an oceanic environment. A description of the situation off Natal has already been given in Chapter 4, and the response of the ocean can be expected to be complex and to depend on a variety of local and remote factors (Gill, 1982).

The amplitude of the barotropic tide varies from about 0.5 m at neap-tides to about 1.75 m at spring-tides, with an indication of an increase with distance northwards. The shelf is too narrow to have a marked influence on this amplitude (Clarke and Battisti, 1981), and tidal currents are generally small compared with those originating in other processes (Schumann and Perrins, 1982). Most analyses place an amphidromic point for the dominant M_2 tide to the south of Africa (e.g

Accad and Pekeris, 1978) with the cotidal lines therefore running nearly parallel to this coast; as a consequence, there is a negligible phase difference between the tides to the north and south of the region.

There are a large number of rivers and estuaries along this stretch of coast. In most cases the discharge of fresh water into the sea tends to be very sporadic, and the influence of this water is local. As a consequence the overall effect is small, as far as the vertical density structures and currents are concerned. Occasionally after heavy rains silt-laden floodwaters penetrate several kilometres out to sea, but such situations are generally short-lived. Nonetheless, in summer waters with slightly lower salinity values are found, and silt can be detected in the circulation patterns further offshore.

MEASUREMENTS

The vast majority of physical oceanographic measurements off Natal have been made by the Physical Oceanography Division of, first the National Physical Research Laboratory, and then the National Research Institute for Oceanology (NRIO). Consequently the description given here will concentrate on the work carried out by this Division, when it was based in Durban during the years 1960 to 1983.

At the start of the investigations in the early 1960s, measurements were confined to a region within a few kilometres of the Natal coast. Current velocities were estimated using drogues tracked from land stations, as well as with dyes whose movement was tracked by aerial photography. Most of these projects were involved with pollution studies, and tended to be concentrated in the vicinity of Durban. The characteristics of the mean flow at different points along the coast were also established from ship-drift estimates (Harris, 1964).

Later, whalers were hired from an industry that was running down, and equipment was developed for measuring temperature profiles. Salinity was determined from water samples taken at specific depths, while currents were measured relative to the ship; accurate navigation was required for determining absolute currents.

With the acquisition of the RV Meiring Naudé (see Appendix) in 1968, more extensive and regular measurement programmes were started. Pollution studies continued, and in fact one of the first tasks that the ship had to carry out, in June, 1968, concerned an investigation of the pollution threat from the 46 000 ton tanker, the World Glory, which broke in half north-east of Durban. The regular cruises off Richards Bay in the years 1970 to 1972 were also designed specifically to determine the feasibility of disposal to sea of effluents from the new harbour and from industrial and housing developments. At the same time, measurements of temperature, salinity and occasionally currents were started on a line of stations offshore from Durban, and at a later stage to the south off Port Edward.

A sophisticated computer-controlled data acquisition system was developed for the RV Meiring Naudé (Stavropoulos, 1971; Snyman, 1980); this also allowed rapid processing of the data obtained. An improved version of the NRIO-built hydrosonde was used, and accurate navigation enabled precise values of currents to be obtained. Water samples were obtained for chemical and nutrient analyses (see Chapters 6 and 7).

It was recognised at an early stage that variations in sea-surface temperature (SST) could be used to derive circulation patterns. In particular, the marked thermal gradient at the inshore edge of the Agulhas Current allowed the variability of the current to be monitored. A Barnes airborne radiation thermometer (ART), model PRT 14-313, was used to map these changes in SST from an aircraft flying at a height of 150 m. This system started in the late 1960s, was used for a number of years, and provided a valuable synoptic-scale picture of ocean structures to complement work done by the ship.

Nonetheless, it soon became clear that time series data would greatly improve the understanding of dynamic processes in the shelf region. A start was therefore made in 1972/73 with the purchase of suitable self-recording meters. The reliable and well-proven Aanderaa RCM-4 meter was chosen, and a suitable data-processing procedure established. As a consequence, a shelf dynamics project was started in 1975.

The main thrust of this project occurred during the period April, 1976, to March, 1977, when moorings were maintained for most of the time at four points along the coast, namely off Richards Bay, Durban, Southbroom and Port Edward (see Figure 5.1). These were primarily deployed on rigid stands in about 30 m of water (Schumann, 1981). Later

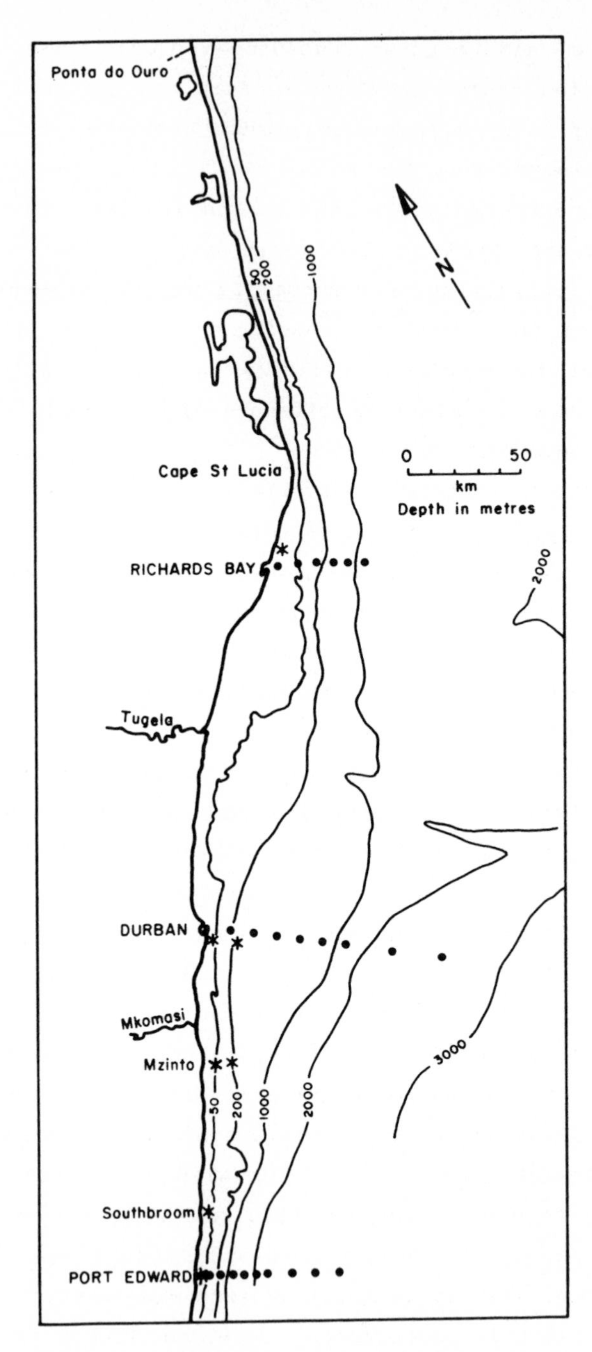

Figure 5.1 Coastal bathymetry and measurement sites off Natal. The lines of dots indicate ship station sites where data were collected regularly; the asterisks show mooring positions reported on in the text.

moorings were deployed in deeper water at specifically selected sites, but were recovered within about two months.

With the acknowledged importance of wind forcing, automatic weather stations were erected at one or two sites to complement the generally unsatisfactory data obtained from Louis Botha airport near Durban (see Chapter 4). None of these stations was maintained for long.

The development of satellite remote-sensing techniques has also been of considerable help in elucidating flow structures. The thermal infrared sensors on board the TIROS and NOAA satellites have given extensive coverage of the ocean areas, while METEOSAT provides pictures more regularly though with much poorer resolution. Images in other spectral bands on LANDSAT and NIMBUS 7 have provided valuable additional information; the fact that South Africa was a member of the Nimbus Experimental Team aided considerably in obtaining these images.

WATER CHARACTERISTICS

The main properties characterizing water type in physical oceanography are temperature and salinity, and many source areas can be identified from so-called TS values. Figure 5.2 shows the distribution of values that can be expected off the Natal coast.

For the shelf areas, only the upper section of the distribution envelope needs to be considered. This shows that these waters generally have a tropical or subtropical origin, with the Agulhas Current system playing an important part in transferring the water to this area. On the other hand, colder, deeper water is occasionally brought onto the shelf by various upwelling mechanisms, particularly off Southern Natal; this is likely to be central water.

Subtropical surface water (STSW) is characterized by a relatively high salinity (greater than 35.5×10^{-3}), caused by greater evaporation. Moreover, it is clear that there are also marked seasonal changes in the upper reaches of the ocean, shown here by the mid-summer envelope. Higher temperatures are due to the increase in summer insolation, with lower salinities due to increased rainfall and outflow from large rivers such as the Zambesi and Limpopo.

Figure 5.3 shows the variabilities in temperature and salinity in

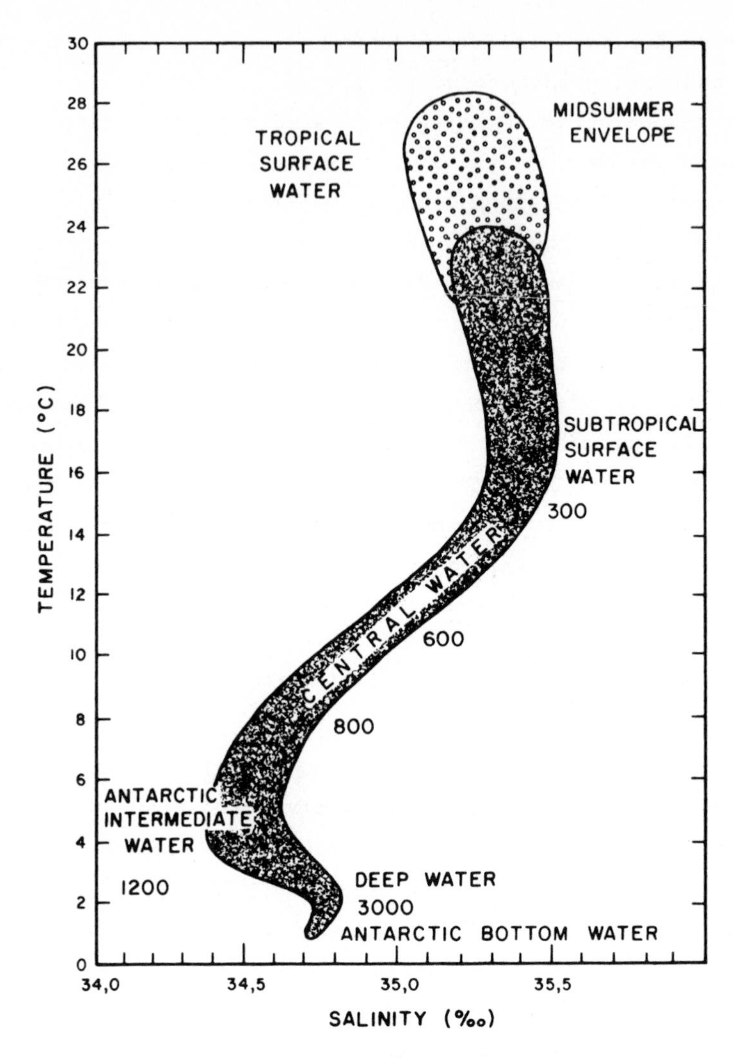

Figure 5.2 A temperature/salinity distribution of values measured off the southeast coast of Southern Africa. Water types are given, along with approximate depths (in m) where such water is commonly found (from Schumann and Orren, 1980).

the water column over a period of about 2 years on the inner and outer shelf off Richards Bay. These sections were chosen to demonstrate the difference between conditions on the shelf and those further offshore on the inner boundary of the Agulhas Current.

The seasonal variability is clearly evident. In the shallower, inshore regions there is about a 4°C temperature change from summer to

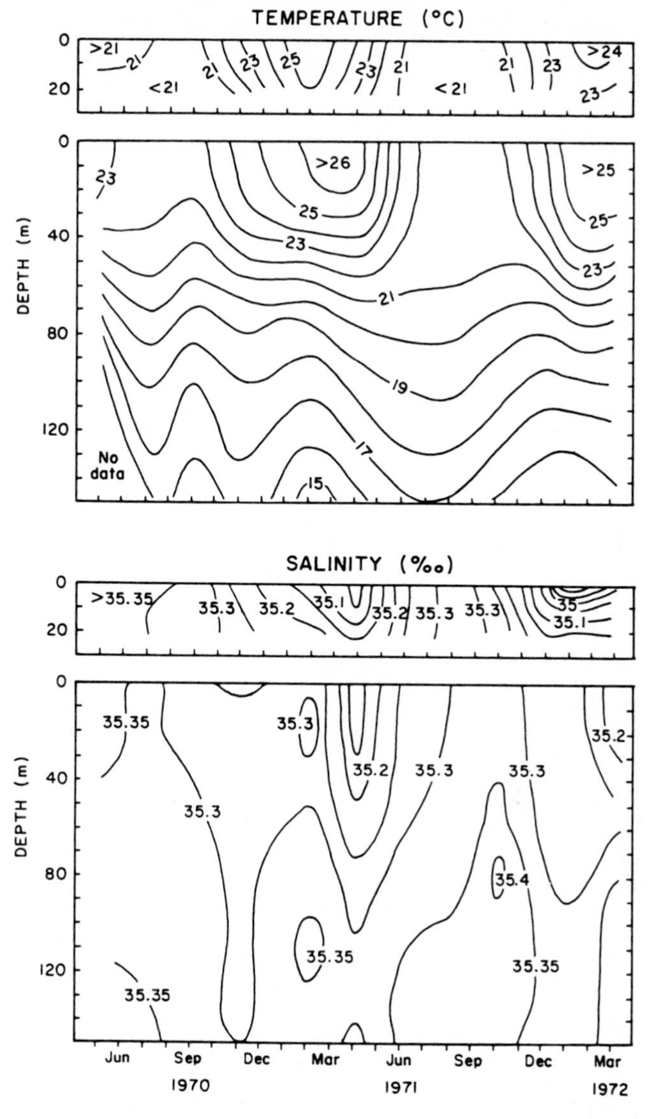

Figure 5.3 Time sections of monthly mean temperatures and salinities off Richards Bay over the two years 1970 to 1972. The shallow, upper sections refer to the inner shelf and the deeper sections to the outer shelf (adapted from Pearce, 1977a)

winter, with maximum temperatures of about 25°C in February. There is also a fairly uniform vertical temperature structure, indicating a well-mixed regime. The most apparent seasonal variation in salinity comes from the relatively fresh water in the upper reaches in late summer, probably as a result of river outflow.

Further offshore there is also about a 4°C temperature change from summer to winter in the upper 50 m of the water column, with maximum temperatures greater than 26°C; deeper down, this seasonal variation is not evident. There is no clear indication of a marked thermocline at any stage, with a regular decrease in temperature with increasing depth. The lowest temperature at 150 m was less than 15°C, and it is of note that this occurred in summer. Salinity variations are not as conspicuous, although there is evidence of water of lower salinity in the upper reaches in late summer.

Analysing the same data, Pearce (1978) found an annual temperature range of 4.8°C, with the inshore water being some 1.4°C cooler than that at the shelf break. He also postulated that water in the 40 to 60 m layer at the shelf break moves shorewards along the appropriate sigma-t surface in a mild but continuous upwelling process.

Whereas the mean seasonal temperature variation along the Natal coast can therefore be expected to be about 5°C, marked changes can occur on a much shorter time scale. Figure 5.4 shows time series of temperature data taken at moorings deployed at three different sites. The seasonal variation displayed in Figure 5.3 is again evident, while there is also no clear longshore trend. However, the abrupt changes which occurred on a day-to-day time scale in many cases exceeded the seasonal variation. These short-term fluctuations can be up to 8°C or 9°C, and are probably caused by the colder deeper water also shown in Figure 5.3 moving up or down the shelf slope and over the measuring site.

THE AGULHAS CURRENT

After its formation in the north, the poleward-flowing Agulhas Current influences the coastal ocean along the whole Natal coast. Its characteristics have been best monitored in the south, and these results will be used to identify aspects to be expected over the whole region. Differences along sections of the coast can be surmised from the measurements available.

Pearce (1977a,b) analysed data from measurements made in the upper 500 m of the Current off Durban, and Figure 5.5 shows mean profiles and

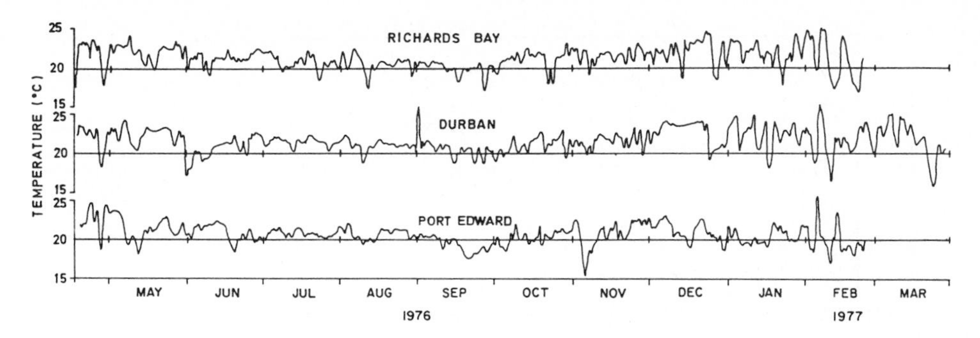

Figure 5.4 Time series of temperatures measured 3 m above the sea bed in about 30 m of water, off Richards Bay, Durban and Port Edward (from Schumann, 1981, JGR, 86 copyright by the American Geophysical Union).

sections from about 45 completed lines of such stations in all seasons. Here the core of the Current lies at a distance of between 40 km and 60 km offshore, with an intense cyclonic shear region on the inshore side, while speeds on the offshore boundary tail off far more gradually.

Sea-surface temperature profiles correlate well with the current speeds, showing the warmer water associated with the Agulhas Current. On the other hand, there is no consistent salinity change on the inshore boundary, with a gradual increase offshore being evident. If the "edges" of the current are defined as the 0.4 m/s isotachs, Bang and Pearce (1978) found the mean width of the Current to be about 100 km.

The subsurface temperature and salinity structures are typical of western-boundary currents, with the slopes of the isotherms and isohalines reaching maxima at the Current core. An intrusion of high-salinity STSW can also be detected at a depth of 150 m to 200 m offshore of this core. Closer inshore, the slopes of the isotherms and isohalines change, and this is associated with a current reversal.

Using data mainly from sections off Port Edward, Gründlingh (1980) found a volume transport for the Agulhas Current of around 70×10^6 m^3/s, extrapolated over the total depth and width. This value is comparable to those obtained for the Gulf Stream (Knauss, 1969) and the Kuroshio Current (Taft, 1972), and confirms that the Agulhas Current is one of the major currents of the world's oceans. No distinct seasonal variation was found in the volume flow, and a more

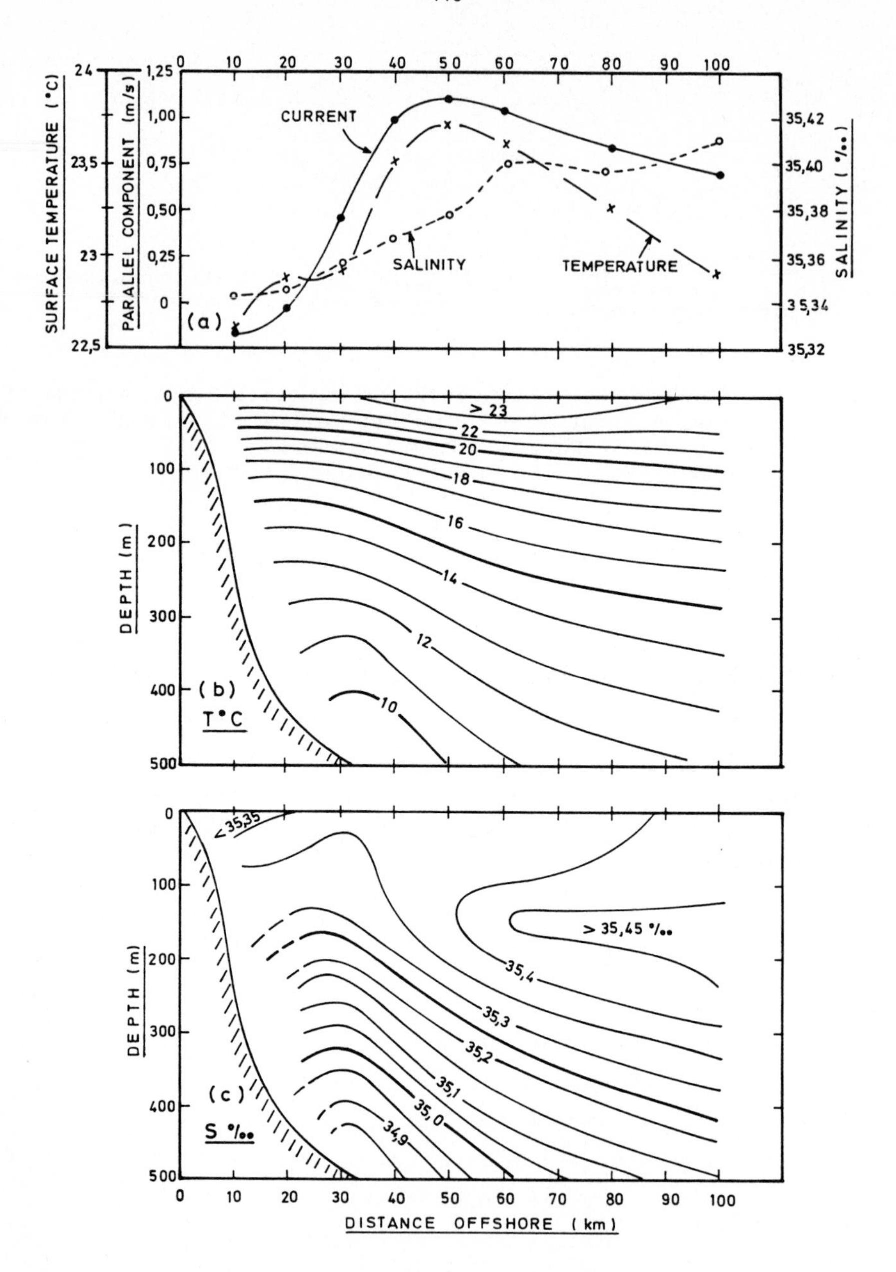

Figure 5.5 Profiles and sections measured on the line of stations off Durban. In (a) are shown the surface profiles of mean temperature, salinity and 0-to-100 m depth-averaged longshore current component, while (b) and (c) depict temperature and salinity, respectively (from Pearce, 1977a).

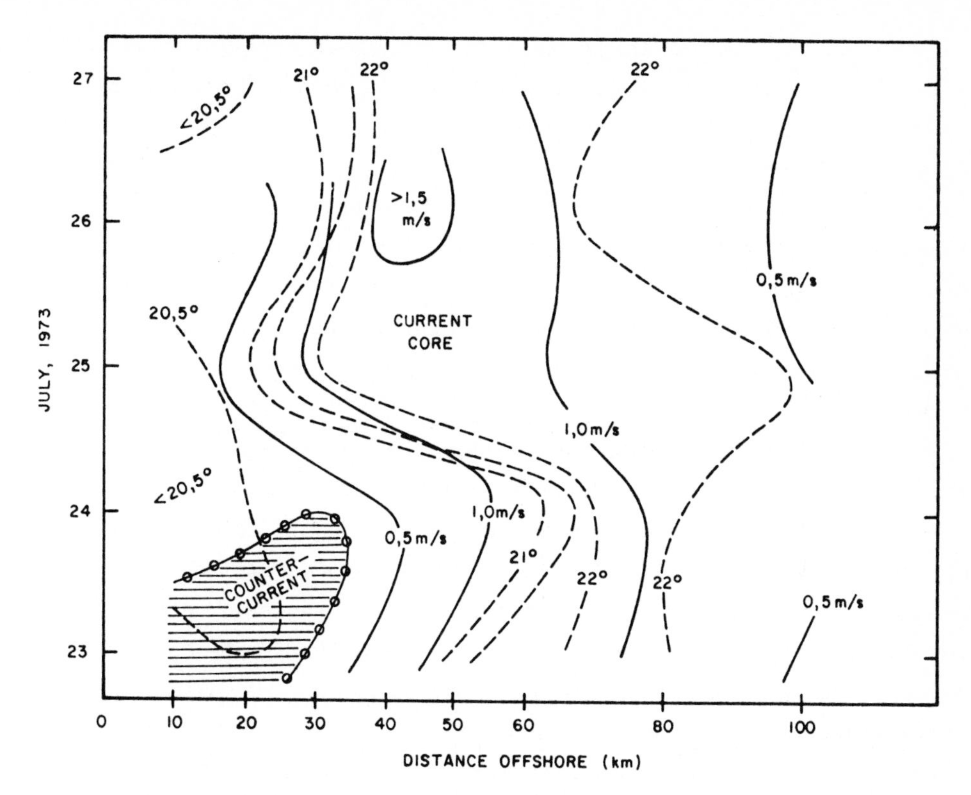

Figure 5.6 A "space-time" map of sea surface temperatures (dashed) and longshore current component during five days of a cruise in July, 1973. The shaded area represents a northeastward counterflow (from Pearce, 1977b).

recent analysis of current speeds (Pearce and Gründlingh, 1982) has confirmed this conclusion, within the limits imposed by the data available.

In fact, it has been found that the daily variability in the Current is such as to mask any seasonal variability that may be present. Figure 5.6 shows the changes measured over a period of 5 days on the Durban line of stations, and demonstrates that the current core can change its position by 30 km and more from day to day. Ostensibly such changes are associated with meanders, and Gründlingh (1974) postulated that these could be initiated by atmospheric forcing.

As mentioned before, it is likely that bottom topography plays an important part in the structure and flow of the Agulhas Current. Off northern Natal there is a well-formed shelf break, and the few measurements available indicate that the Current usually flows just offshore of this break. South of Cape St Lucia a gentle bight in the coastline causes the Current to move offshore, while the terrace-like topography off Durban affects the structure below about 500 m. To the south the shelf narrows considerably, and observations show an increase in Current speed relative to Durban, with the core situated within about 10 km of the shore off Port Edward. Gill and Schumann (1979) modelled these changes in terms of topographic control of the Current flow.

These features are shown in the sea surface temperatures depicted in Figure 5.7. The Agulhas Current boundary is clearly evident from the strong horizontal temperature gradient, with the apparent onset of a meander south of Cape St Lucia. There is a dramatic growth in the meander off Durban, with a strong onshore tendency off Mzinto and indications of a recirculation flow. Further south off Port Edward the Current has apparently become re-established. A mean sea surface temperature figure obtained by Pearce (1977b) in 14 ART flights, shows a distinct 'kink' in the isotherms off Durban. It is therefore apparent that the meander in Figure 5.7 bears a specific relationship to its position over the bottom topography.

A further feature which has been observed in satellite imagery of this area is the so-called "Natal pulse". These pulses consist of large deflections of the core of the Agulhas Current from its mean position, originating off Natal and growing with distance downstream to beyond Port Elizabeth (Lutjeharms, 1981; Gründlingh, 1979).

THE NATAL SHELF

The current-meter moorings maintained at Richards Bay, Durban, Southbroom and Port Edward during 1976 and 1977 were intended to investigate longshore coherence scales. However, the analysis of subtidal fluctuations carried out by Schumann (1981) essentially showed a lack of coherence between conditions at the sites selected, and the existence of distinct coastal regimes. It was postulated that these different regimes occurred as a result of changes in bottom topography

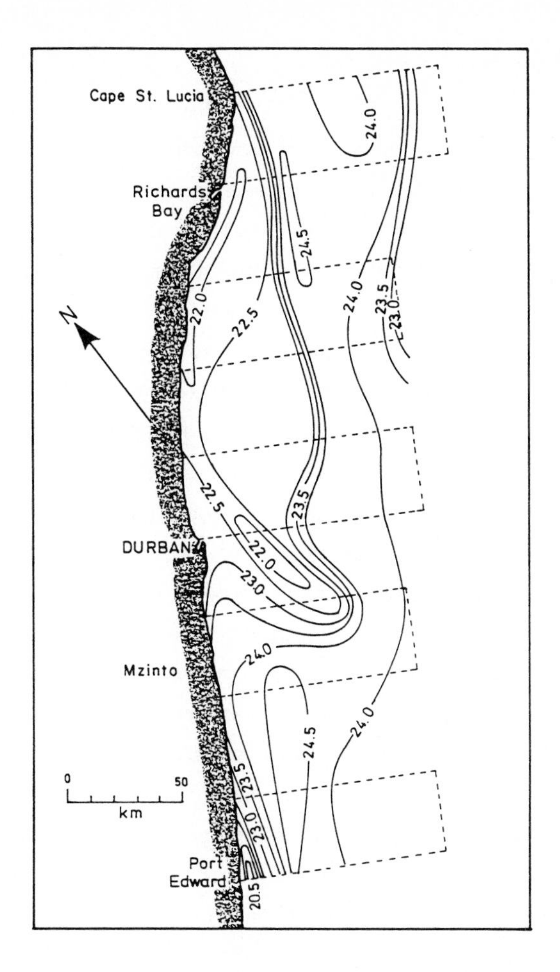

Figure 5.7 Sea surface temperatures from an ART flight on 30 November, 1967. Temperatures are given in degrees Celsius, while the aircraft flight pattern is also shown (from Schumann, 1982).

and also by the varying influence of the Agulhas Current at different sections of the coast.

Four broad regimes are identified, and will be discussed separately. It must be clearly understood that such a subdivision is by no means absolute, but that it merely serves to identify areas with generally different dynamic characteristics. There are no clearly defined boundaries and, indeed, if more comprehensive data were available different subdivisions might be made.

The Northern Region

This is roughly the region north of Cape St Lucia, that is the coast of Maputaland, where the shelf is narrower than 10 km in places (Figure 5.1). Virtually no inshore measurements are available for this region, but it can be surmised that the Agulhas Current will have a dominant influence, particularly where it flows close to the shelf break.

North of Cape St Lucia maximum speeds of about 1.5 m/s have been measured in the Current. There does not appear to be such a distinct thermal gradient on the inshore boundary as there is further south, which perhaps indicates that the Current is still in its formative stages. The slight change in the direction of the coastline leading to the bay at Maputo and the major coastal inflexion northeastwards (Figure 1.1), means that the main flow of the Agulhas Current may also lie further offshore. Satellite imagery has revealed the existence of an eddy situated offshore of Maputo, also depicted in the analysis by Saetre and Da Silva (1984). It is therefore probable that the shelf regions in the far north are not as strongly under the influence of the Agulhas Current as they are further south.

These conclusions are also supported by Harris (1964, 1978), in analyses of data of ship-drift from positions about 2 km offshore. Thus it was found that the percentage of northeastward currents increased to the north. However, northeastward current speeds tended to be less than 0.25 m/s (0.5 knots), while those of southwestward currents generally exceeded 0.5 m/s (1 knot).

The Central Shelf

South of Cape St Lucia the shelf becomes wider, mainly due to a gentle bight in the coastline, and the main stream of the Agulhas Current also diverges from the coast. The widest part of the Natal shelf is off the Tugela River, where the shelf break is about 45 km offshore. It is this wider shelf region which stretches southwards to the start of the terrace-like structure north of Durban that will be discussed here.

Results from the hydrographic measurements made off Richards Bay show the marked effect of the Current at the shelf break and beyond (Pearce,

1977a). The results obtained from the mooring closer inshore and just to the north of Richards Bay (Figure 5.1), indicate the shelf-like characteristics that can be expected further away from the influence of the Current. Thus it is apparent that there are two processes operating: first, in the several-day period range the current fluctuations lie parallel to the coastline, and second, the effect of the Agulhas Current is evident in the longer-term current progression southwestwards along the line of the deeper contours (Schumann, 1981).

In seeking an explanation for the shorter-term fluctuations, the wind was the obvious choice. Figure 5.8 shows time series of wind and current vectors for a portion of the time for which current data are available. It is clear that there is a very close correspondence, and coherency and phase spectra between the longshore wind and longshore and offshore currents confirm this conclusion (Schumann, 1981). A markedly linear relationship was found between phase and frequency, giving a time lag of about 18 hours at all frequencies between the onset of the wind and the response of the currents; this is within the inertial period of about 25 hours. Previously, Bang and Pearce (1978) had identified current fluctuations in the 4- to 6-day period range, but ascribed this to an association with atmospheric pressure lows.

In his analysis of the hydrographic measurements off Richards Bay, Pearce (1977a) found that there was no major seasonal pattern in the behaviour of shelf currents. Any such pattern was probably obscured by the high short-term variability caused by the wind-driven currents inshore, and by the meandering of the Agulhas Current at the shelf break. As mentioned before, the mean flow structures indicated that the cross-shelf flow appeared to be onshore in the lower half of the water column, with an offshore compensatory flow in the upper layer; this was ascribed to Ekman veering in the bottom boundary layer of the Agulhas Current.

Pearce (1977a) also found sporadic instances of localised upwelling, typically within about 10 km of the coast and some 20 km in alongshore extent. Surface temperature dropped by about 1°C and higher levels of nutrients were found in such cases (Chapter 6). Such events were shortlived, however, and seldom lasted much longer than one day.

Very few measurements are available on the wider shelf to the south, and it must be assumed that many of the characteristics of the inner

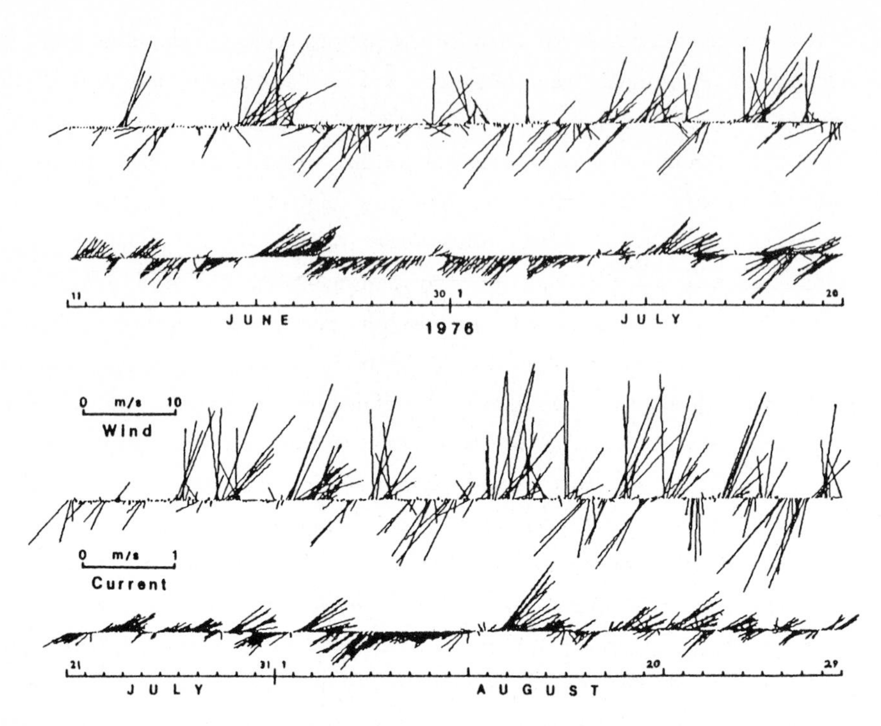

Figure 5.8 Vector time series of winds and currents measured at Louis Botha airport near Durban and Richards Bay, respectively. Filtered, 3-hourly values have been used, and the 80 days of data have been broken into two sections with the wind vectors at the top in each case. North is at the top of the page, while wind and current scales are also given (from Schumann, 1981, JGR, 86, copyright by the American Geophysical Union).

shelf off Richards Bay also apply here. The analysis of ship drift data by Harris (1964) confirms the variability of the currents in the area, with an almost equal percentage of northeastward and southwestward currents. There is an indication of a higher percentage of northeastward currents in the south, although the limited number of measurements available precludes any definite conclusions to be drawn in this regard.

A further source of information on current patterns is satellite imagery. Figure 5.9 shows an enhanced LANDSAT MSS band 4 image, in which the absorption of radiation by particulate-free ocean water is minimal and penetration can be 10 m or more. Clear water is shown as black, while water containing suspended particulate matter (river-borne sediment, industrial effluent, plankton, etc.) is shown in various

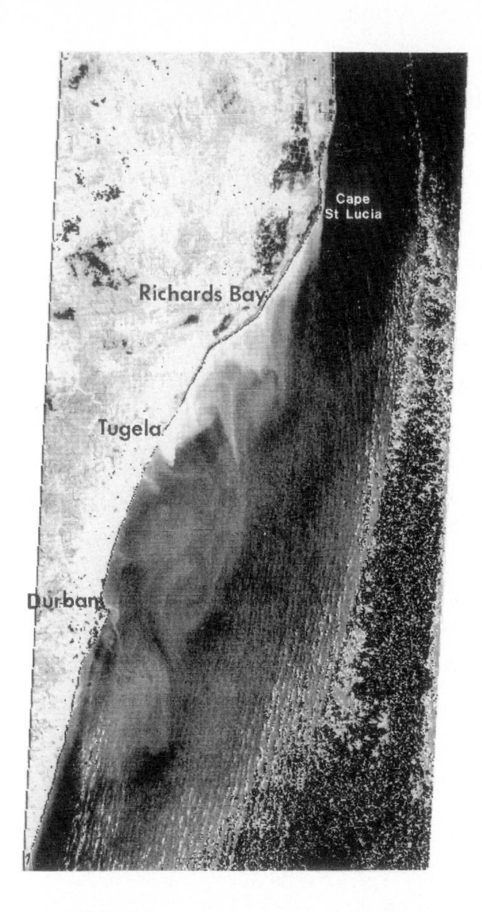

Figure 5.9 A computer-enhanced picture of the LANDSAT-1 multispectral scanner band 4 (0.5 to 0.6 μm) image acquired on 29 January, 1973 (from Malan and Schumann, 1979).

shades of grey, depending on the concentration and depth of the suspended matter. Rainfall reports at the time indicate that there would have been substantial discharges from the major rivers in the area in the days preceding the image.

The distribution of particulate matter in the water reveals a complex pattern of flow over the Natal shelf. The "average" circulation situation described by Pearce et al. (1979) is supported, with the existence of eddies or similar features with a variety of scales. The most persistent feature is the cyclonic eddy immediately offshore and to the north of Durban, while to the north-east of the Tugela are southward-extending plumes with an indication of the presence of a smaller eddy. Closer to the coast there are smaller-scale features, probably associated with the varying current regime.

The Durban Shelf

This is a region extending from an ill-defined area to the north of Durban, to a point south of Mzinto where the shelf is narrower. It may be thought of as a transition region between the wind-dominated shelf to the north, and the Agulhas Current dominated shelf to the south. However, there appear to be some very specific characteristics which serve to identify this as more than a transition region, while there is also little long-term coherence with the other two regions (Schumann, 1981).

Figure 5.10 shows current vectors measured on the inner stations of the lines off Durban and Port Edward. One of the main results to emerge is the inherent variability of the currents off Durban. The trend from a generally northeastward flow at the innermost station to a southwestward flow at about 50 km offshore is evident, although even at this latter site very slack current conditions were encountered at times.

With the Agulhas Current flowing southwestward further offshore, a source for the northeastward inshore flow probably has to be sought in a recirculation of waters on the inshore boundary of the Current. Such a flow pattern is indicated in Figure 5.7, and is given strong support by simultaneous current meter results from moorings off Durban, Mzinto and Port Edward (Schumann, 1982).

These results reveal a predominantly northeastward flow off Durban, though with the occurrence of fairly regular current reversals. Off Port Edward the flow is strongly southwestward, but with an onshore flow over the whole water column (195 m) off Mzinto. The consistent nature of this flow is the best evidence to date of the fairly permanent nature of the recirculation of the inshore waters of the Current. The temperature records also support this recirculation concept, in that the temperatures measured at the upper meter at Mzinto, and those measured at the inner mooring off Durban generally remained within about 0.5°C of each other over the whole 51-day recording period.

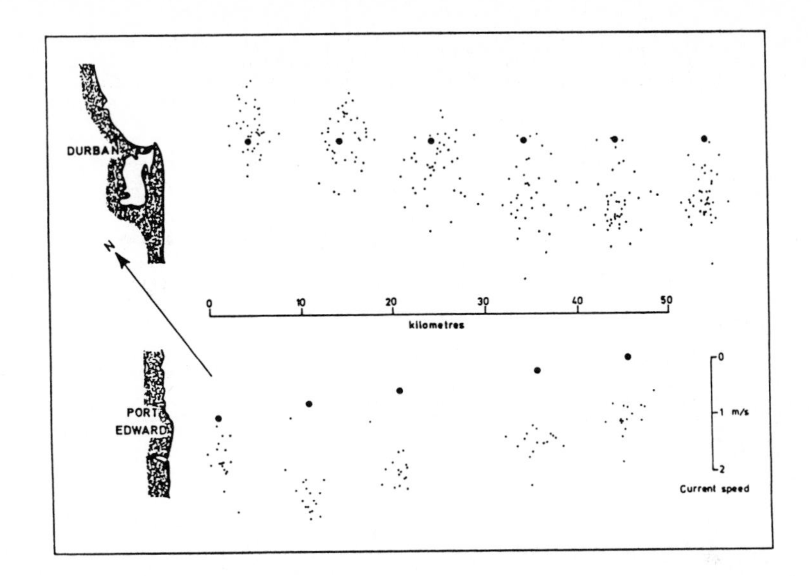

Figure 5.10 Current measurements made on the line of stations off Durban and Port Edward. Each dot represents the head of a current vector averaged over the upper 100 m with respect to the associated station position, represented by the bigger dots (from Schumann, 1982)

The nature of the current fluctuations is also of interest. Figure 5.11 shows current and temperature records taken on another occasion over a period of about 34 days off Durban and Mzinto. The essentially barotropic current structures are apparent, with the variations over the whole water column taking place simultaneously at both sites off Durban.

Wind and air pressure records also reveal little correlation, and it is therefore likely that the cause of such current fluctuations lies in the Agulhas Current variability and changes in the recirculation flow.

The temperature records show that the measurements at the deeper site off Durban were essentially taken below the thermocline. At the other two shallower sites the temperature changes are much more dramatic, indicative again of colder bottom water reaching the measurement site at times. At Mzinto the current speeds were much lower than off Durban with a greater variability being evident. However, of particular note is the close correlation between the temperature variations and the current fluctuations at Durban and Mzinto. Thus with a northeastward

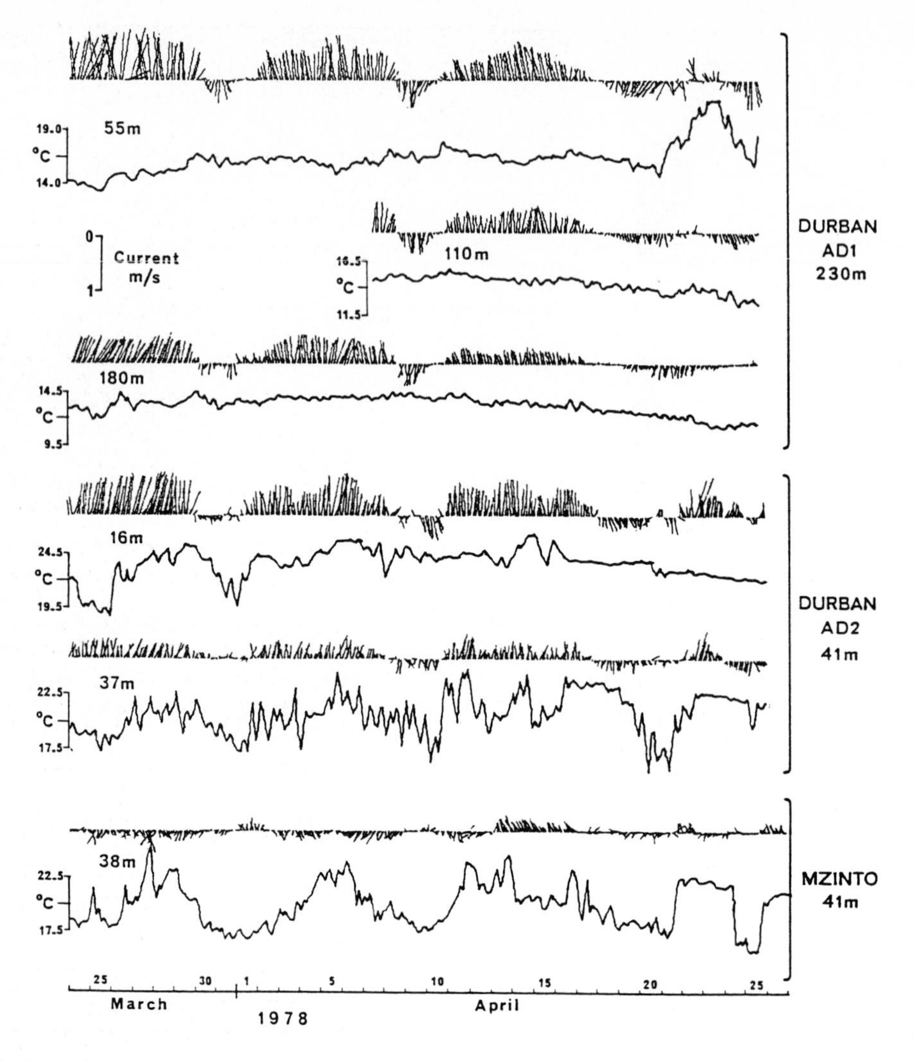

Figure 5.11 Current and temperature time series from moorings off Durban and Mzinto during 1978. The total depth at each mooring is given, as well as the depth of the individual meters. The orientation of the coastline is at 38°T, here parallel to the page margin (from Schumann, 1982)

flow the temperature reached a maximum of about 22.5°C, while a southwestward flow accompanied a drop of around 5°C. The correspondence in terms of actual values and variations in the bottom two temperature records also confirm the existence of connections between the two sites.

On two occasions drogued buoys were released and followed for a period of about 36 hours (Schumann, 1982, with acknowledgement to R. A Carter). One, starting from a position about 20 km offshore, followed roughly the pattern indicated by the gyre offshore of Durban, shown in Figure 5.10, and this may be the northward tendency of the recirculation flow. On the other hand, the second buoy released 10 km offshore first went southwards before turning northwards.

Closer inshore the break in the coastline formed by the Bluff at Durban is apparently responsible for causing a southward flow close to the coast in the small bight. This was shown up by early drogue measurements, and has been ascribed to an anticyclonic gyre formed inshore of the main northward flow (Pearce, 1977a). The rest of the Natal coastline is relatively straight, and it is unlikely that there are other substantial local topographically-related flows.

A number of drogue measurements were also made in the early 1960s. within 5 km of the coast off the Mkomasi river. While the currents recorded were generally very variable, it is clear that on occasion the influence of the Agulhas Current was felt strongly, with southwestward flows exceeding 1 m/s.

The Southern Region

This is a region becoming increasingly under the influence of the Agulhas Current. This is evident from the current vectors depicted in Figure 5.10, which show that current speeds exceeding 1 m/s often occur within 10 km of the coast. Sections offshore of Port Edward show the Current core, with steeply sloping isopycnals, just offshore of the shelf break.

Figure 5.12 shows progressive vector diagrams (PVDs) from meters deployed at roughly the same distances offshore at Southbroom and Port Edward (Figure 5.1). The strong southwestward flow is evident, but it is also markedly stronger at the latter site; this can probably be ascribed to the still narrowing shelf and to the consequent onshore movement of the Current to the south.

Nonetheless, it is also clear that current reversals do occur on occasion, and it is likely that these are associated with variations in the Agulhas Current itself. A coherency analysis between currents

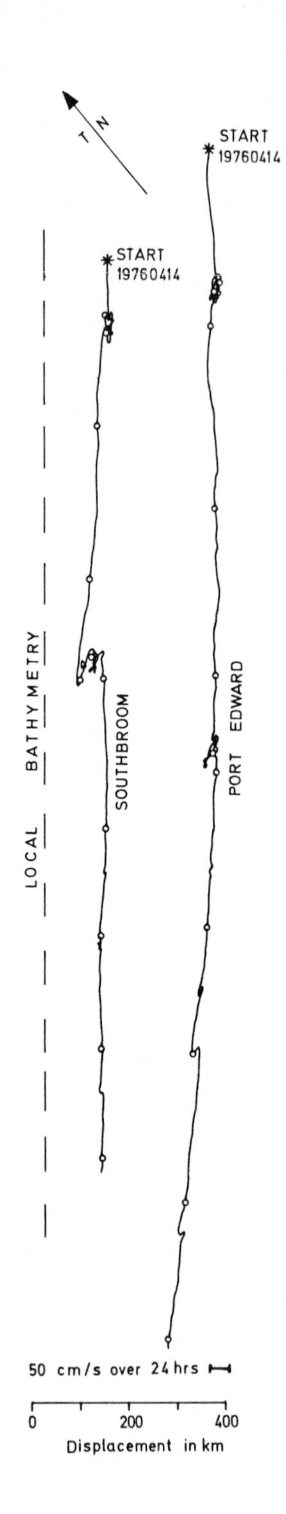

Figure 5.12 Progressive vector diagrams (PVDs) of current measurements covering some 110 days, taken in about 30 m depth at Port Edward and Southbroom. The starting days are given by an asterisk, with an open circle every 10 days. The orientation of the bathymetry and coastline are also shown.

and the longshore winds showed that local wind forcing plays a relatively minor role in this area (Schumann, 1981).

The strong currents on the shelf mean that substantial bottom boundary layer effects can be expected. A recent analysis by Schumann (1986) in 49 m water depth showed that the turbulent flow associated with this layer extended over the whole water column. Ekman veering in excess of 15° was found, and this serves to bring colder, deeper water onto the shelf. Such upwelling does indeed occur along this coast, and Figure 5.7 gives an indication of this off Port Edward. More substantial upwelling is common further south, but it is not clear which mechanism is responsible: either Ekman veering in the bottom boundary layer, or perhaps the topographic mechanism described by Gill and Schumann (1979).

DISCUSSION

With the acknowledged economic and social importance of the coastal oceans of the world, considerable effort has been expended in efforts to understand the processes operating there. However, it is clear from the outset that there can be marked differences between coastal regions. The geographic setting plays an important part, with notable variations between the east and west coasts of the continents.

The orientation of the coastline, latitude, the extent of shelf area and disposition of neighbouring land masses are additional complicating features, quite apart from differing weather patterns. In order to assess the relevance of the results for the Natal coast in relation to those from other regions, it is therefore necessary to identify similar components.

The dominant feature is the western boundary current, which immediately identifies an east coast environment. Such corresponding currents occur off the east coasts of Japan, Australia, and North and South America. Of these, by far the most investigations have been made off the southeast United States, where the Gulf Stream flows polewards adjacent to the shelf.

The South Atlantic Bight (SAB) region, extending over about 9° of latitude, has points of similarity with the Natal coast. However, the shelf there is wider, varying from about 50 km to 120 km, with the shelf break occurring at about the 75 m contour. On this shallow shelf, Atkinson _et al_. (1983) identified three depth-dependent zones: 0-20 m, 21-40 m and 41-60 m. The inner shelf responds quickly to atmospheric conditions, and temperatures in summer exceed 28°C; salinity is also affected markedly by river outflow. At the outer shelf most of the variability is due to Gulf Stream frontal effects which occur at periods of 2 days to 2 weeks (Lee _et al._ 1981), but which are more akin to the eddies observed in the Agulhas Current east of the Agulhas Bank (Lutjeharms, 1981) than to the meanders observed off Natal.

Lee and Atkinson (1983) analysed the low-frequency variability over the SAB shelf using results from an extensive set of moorings. They found that along-shelf current fluctuations at the 40 m isobath were highly coherent with the along-shelf wind and coastal sea level at periods of 3 to 4 days and 10 to 12 days. These current fluctuations were in phase over the whole length of shelf, indicating a non-propagating barotropic response to wind-driven sea level fluctuations. At the shelf edge Gulf Stream features were observed to propagate northwards at speeds of 0.5 to 0.7 m/s, producing fluctuations of currents and temperatures on the shelf with along-shelf coherence scales of 100 km. These results have aspects which are similar to those obtained on the Natal coast, but it is obvious that the SAB does not have the same along-shelf variability.

Coastal trapped waves have also been found to play an important part in the dynamics of many coastal regions (Le Blond and Mysak, 1978). Lee and Atkinson (1983) did not observe such southward propagating waves in the SAB, although Schott and Düing (1976) found significant evidence for the propagation of trapped waves further south in the Florida Straits.

Schumann (1983) analysed available data but found no clear evidence for the existence of coastal trapped waves off the southeast coast of Southern Africa. This was probably due in part to the use of data from the differing regimes off Natal, although further south conditions may be more favourable. Nonetheless, he also described an event propagating northwards at about 3.4 m/s between Port Edward and Richards Bay; the temperature record in Figure 5.4 shows this signal during April, 1976. It is of interest to note that Brooks and Mooers (1977) found that the coherence between sea level stations decreased rapidly when moving north of the Florida Straits into a widening shelf region. As off Natal, this would also seem to imply that the differing shelf structure and Gulf Stream inhibited the propagation of shelf waves.

The situation is different off the east coast of Australia. There the East Australian Current (EAC) separates from the coast at about 31°S or 32°S, and flows eastwards into the Pacific Ocean (see e.g. Cresswell _et al_. 1983). Complex eddies result, but the EAC does not have the same effect on the dynamics of the coastal shelf as does the Agulhas Current off Natal. Hamon (1976) also established that coastal trapped waves were a regular wind-forced feature of the East Australian coast.

The bimodal structure of the Kuroshio (Taft, 1972) and the shelf and island structure off Japan point to a complex dynamical region. Nonetheless, Shoji (1961) was able to follow the progression of what was later discovered to be a shelf wave along the Japanese east coast.

Off South America the Brazil Current flows southwards nearly parallel to the edge of the continental shelf between 32°S and 38°S, before turning southeastwards towards deeper water (Legeckis and Gordon, 1982). Further south a confluence occurs with the northward-flowing Falklands (Malvinas) Current. Conditions on the continental shelf areas are poorly understood at present.

It is clear that, while all these east coast regions have some similarities, there are specific characteristics which serve to identify each one uniquely. The region discussed here, namely, the coastal ocean off Natal, is also only one part of the south-east coast of Southern Africa, dominated by the flow of the Agulhas Current.

In conclusion, some satellite images are shown to highlight features discussed earlier. Figure 5.13 is an example of what might be termed the "average" situation; both the thermal infrared band and a processed level 2 image giving subsurface radiances are shown.

The main stream of the Agulhas Current is seen to diverge from the coast south of Cape St Lucia, and then to move inshore again south of Durban with some indication of meandering. Various features are evident on the Natal shelf, highlighted by the presence of total suspended solids of terrigenous biological and non-biological origin (such as silt, clay and other types of sediment). A plume extending out from the Tugela River is clearly evident, with substantial rainfall reported inland about a week earlier. Walters and Schumann (1985) have investigated the detection of silt utilising the NIMBUS-7 coastal zone colour scanner, with other examples along this coast.

On the other hand, Figure 5.14 shows what can only be described as an anomalous situation. Images of sea surface temperature taken on consecutive days reveal a large cyclonic eddy extending well into the region where the Agulhas Current can normally be expected. No additional, concurrent measurements are available, so that it is not clear where the flow of the Agulhas Current was at that time. However, the flow patterns seem to indicate that during such an event waters from the Natal shelf may be entrained into the deeper ocean. As mentioned earlier, such major perturbations on the flow have been

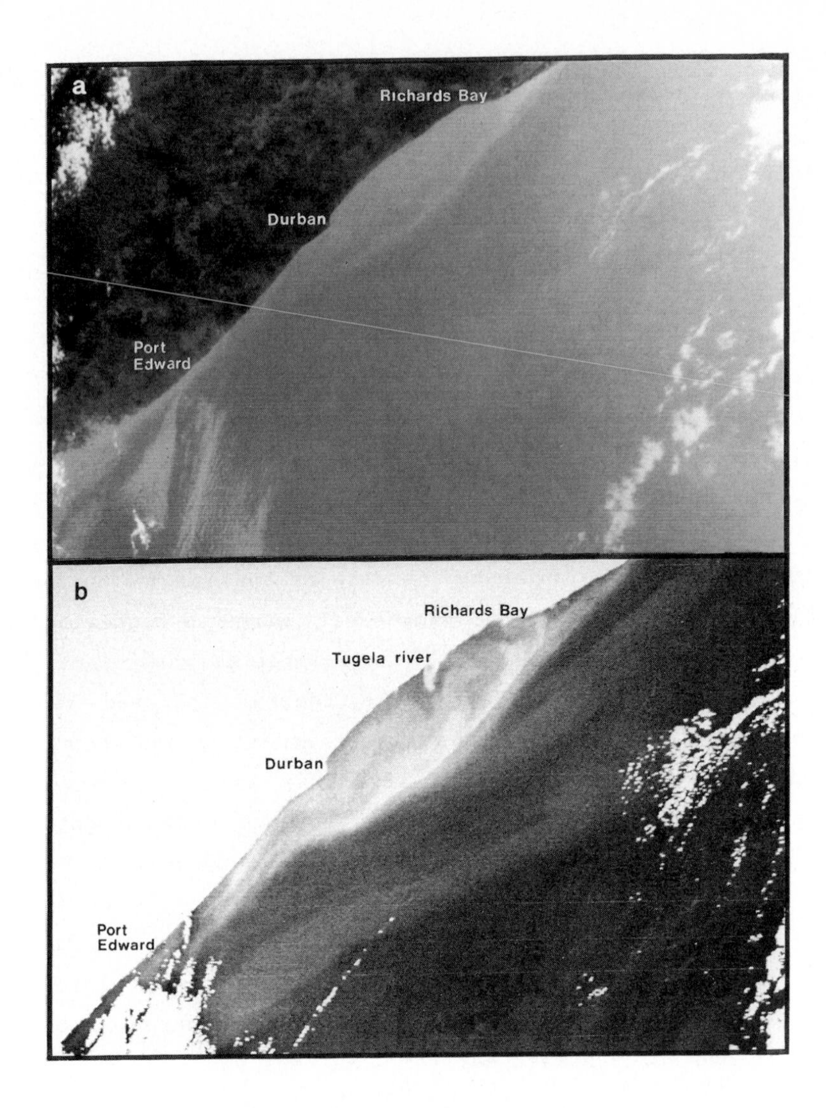

Figure 5.13 Data from the NIMBUS-7 Coastal Zone Colour Scanner acquired on 11 March, 1979. The upper picture (a) shows the channel 6 thermal infrared (11.5 m) image, while below is shown the level 2 image depicting subsurface radiances (with acknowledgement to NASA and the Nimbus Experimental team).

observed from time to time, and Gründlingh and Pearce (1984) also describe the large cyclonic vortex which almost brought oil spilled from the stricken tanker World Glory onto the coast.

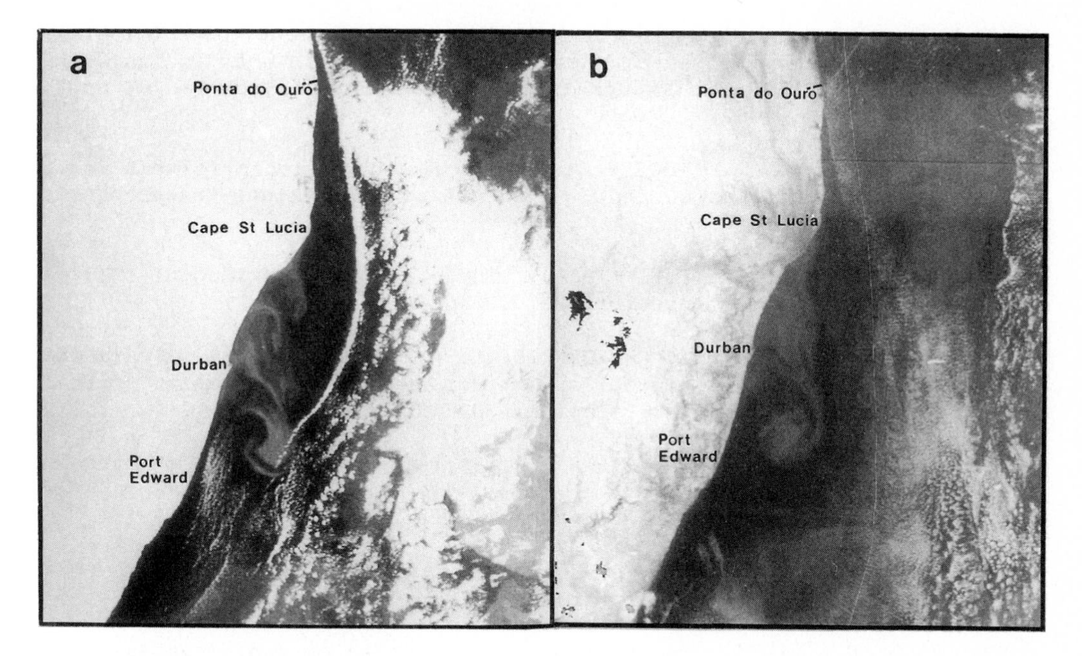

Figure 5.14 Infrared images taken on 21 and 22 April, 1983 (a and b respectively), by the NOAA-7 satellite (with acknowledgement to NOAA).

Acknowledgements

Much of the work was completed while the author was employed by the National Research Institute for Oceanology, and the contribution made by that Institute, and the use of facilities is gratefully acknowledged. In particular, the develpments in data collection and processing made by the late Chris Snyman added substantially to the success of the various programmes.

REFERENCES

ACCAD, Y and C P PEKERIS (1978). Solution of the tidal equations for the M_2 and S_2 tides in the world oceans from a knowledge of the tidal potential alone. **Philosophical Transactions of the Royal Society,** London, Ser. A290, 235-266.

ATKINSON, L P, T N LEE, J O BLANTON and W S CHANDLER (1983). Climatology of the southeastern United States continental shelf waters. **Journal of Geophysical Research,** 88, 4705-4718.

BANG, N D and A F PEARCE (1978). Physical Oceanography. In: Ecology of the Agulhas Current region: an assessment of biological responses to environmental parameters in the south-west Indian Ocean (Ed: AEF HEYDORN) **Transactions of the Royal Society of South Africa,** 43, 156-162.

BROOKS, D A and C N K MOOERS (1977). Wind-forced continental shelf waves in the Florida Current. **Journal of Geophysical Research,** 82, 2569-2576.

CLARKE, A J and D S BATTISTI (1981). The effect of continental shelves on tides. **Deep-Sea Research,** 28, 665-682.

CRESSWELL, G R, C ELLYETT, R LEGECKIS and A F PEARCE (1983). Nearshore features of the East Australian Current system. **Australian Journal of Marine and Freshwater Research,** 34, 105-114.

GILL, A E and E H SCHUMANN (1979). Topographically induced changes in the structure of an inertial coastal jet: application to the Agulhas Current. **Journal of Physical Oceanography,** 9, 975-991.

GILL, A E (1982). **Atmosphere-Ocean Dynamics.** Academic Press, New York.

GRÜNDLINGH, M L (1974). A description of inshore current reversals off Richards Bay based on airborne radiation thermometry. **Deep-Sea Research,** 21, 47-55

GRÜNDLINGH, M L (1979). Observations of a large meander in the Agulhas Current. **Journal of Geophysical Research,** 84, 3776-3778.

GRÜNDLINGH, M L (1980). On the volume transport of the Agulhas Current. **Deep-Sea Research,** 27, 557-563.

GRÜNDLINGH, M L and A F PEARCE (1984). Large vortices in the northern Agulhas Current. **Deep-Sea Research,** 31, 1149-1156.

HAMON, B V (1976). Generation of shelf waves on the East Australian coast by wind stress. **Mémoires Société Royale des Sciences de Liége,** 6e série. 10, 359-367.

HARRIS, T F W (1964). Notes on Natal coastal waters. **South African Journal of Science,** 60, 237-241.

HARRIS, T F W (1978). Review of coastal currents in Southern African waters. **South African National Scientific Programmes Report No. 30,** CSIR, 103 pp.

KNAUSS, J A (1969). A note on the transport of the Gulf Stream. **Deep-Sea Research,** 16, 117-123.

LE BLOND, P H and L A MYSAK (1978). **Waves in the ocean,** Elsevier, New York, 602 pp.

LEE, T N and L P ATKINSON (1983). Low-frequency current and temperature variability from Gulf Stream frontal eddies and atmospheric forcing along the southeast US outer continental shelf. **Journal of Geophysical Research,** 88, 4541-4568.

LEE, T N, L P ATKINSON and R LEGECKIS (1981). Detailed observations of a Gulf Stream frontal eddy on the Georgian continental shelf, April 1977. **Deep-Sea Research,** 28, 347-378.

LEGECKIS, R and A L GORDON (1982). Satellite observations of the Brazil and Falkland currents - 1975 to 1976 and 1978. **Deep-Sea Research,** 29, 375-401.

LUTJEHARMS, J R E (1981). Features of the southern Agulhas Current circulation from satellite remote sensing. **South African Journal of Science,** 77, 231-236.

LUTJEHARMS, J R E, N D BANG and C P DUNCAN (1981). Characteristics of the currents east and south of Madagascar. **Deep-Sea Research,** 28,879-899.

MALAN, O G and E H SCHUMANN (1979). Natal shelf circulation revealed by LANDSAT imagery. **South African Journal of Science,** 75, 136-137.

PEARCE, A F (1977a). The shelf circulation off the east coast of South Africa. **CSIR Research Report No 361.**

PEARCE, A F (1977b). Some features of the upper 500 m of the Agulhas Current. **Journal of Marine Research,** 35, 731-753.

PEARCE, A F (1978). Seasonal variations of temperature and salinity on the northern Natal continental shelf. **South African Geographical Journal,** 60, 135-143.

PEARCE, A F, E H SCHUMANN and G S H LUNDIE (1979). Features of the shelf circulation off the Natal coast. **South African Journal of Science,** 74, 328-331.

PEARCE, A F and M L GRÜNDLINGH (1982). Is there a seasonal variation in the Agulhas Current? **Journal of Marine Research,** 40, 177-184.

SAETRE, R and A J DA SILVA (1984). The circulation of the Mozambique channel. **Deep-Sea Research,** 31, 485-508.

SHOJI (1961). On the variations of the daily mean sea levels along the Japanese islands. **Journal of the Oceanographical Society of Japan,** 17, 141-152.

SCHOTT, F and W DUING (1976). Continental shelf waves in the Florida Straits. **Journal of Physical Oceanography,** 6, 451-460.

SCHUMANN, E H (1981). Low frequency fluctuations off the Natal coast. **Journal of Geophysical Research,** 86, 6499-6508.

SCHUMANN, E H (1982). Inshore circulation of the Agulhas Current off Natal. **Journal of Marine Research,** 40, 43-55.

SCHUMANN, E H (1983). Long-period coastal trapped waves off the southeast coast of Southern Africa. **Continental Shelf Research,** 2, 97-107.

SCHUMANN, E H (1986). The bottom boundary layer inshore of the Agulhas Current off Natal in August, 1975. **South African Journal of Marine Research**, 4, 93-102.

SCHUMANN, E H and M J ORREN (1980). The physico-chemical characteristics of the south-west Indian ocean in relation to Maputaland. In: **Studies on the ecology of Maputaland** (Eds: M N BRUTON and K H COOPER). Rhodes University and the Natal Branch of the Wildlife Society of Southern Africa. 8-11.

SCHUMANN, E H and L-A PERRINS (1982). Tidal and inertial currents around South Africa. **Proceedings of the Eighteeenth Coastal Engineering Conference.** ASCE/Cape Town, South Africa, 2562-2580.

SNYMAN, C G (1980). **An oceanographic data acquisition system using a digital computer.** M.Sc. Thesis, University of Natal, 153 pp.

STAVROPOULOS, C C (1971). Data acquisition on the RV Meiring Naudé. **Electronics and Instrumentation,** May 1981, 11-15.

TAFT, B (1972). Characteristics of the flow of the Kuroshio south of Japan. In: **Kuroshio, Physical aspects of the Japan current** (Eds: STOMMEL and K YOSHIDA), Chapter 6, 164-216, University of Washington Press, Seattle.

WALTERS, N M and E H SCHUMANN (1985). Detection of silt in coastal waters by NIMBUS CZCS. In: **The South African ocean colour and upwelling experiment.** (Ed: L V SHANNON) Sea Fisheries Research Institute, 219-225.

Chapter 6

INORGANIC NUTRIENTS IN NATAL CONTINENTAL SHELF WATERS

Robin Carter and Jeannette d'Aubrey
National Research Institute for Oceanology
Council for Scientific and Industrial Research

INTRODUCTION

The concentrations of the inorganic micronutrients, nitrate-nitrogen, reactive silicate and phosphate-phosphorus, in the sea provide information on fertility (i.e. potential for algal growth), the biological history of the water and water mass identification. These parameters have thus received extensive attention in marine studies in general and in biological studies, e.g. plankton dynamics, in particular.

Natal continental shelf waters are generally well mixed and uniform in their temperature/salinity characteristics. This has led to problems in water mass identification and it was initially for this reason that nutrient concentrations were measured. With the expansion of plankton research in the seas bordering Natal the growth-controlling function of nutrients became apparent and the measurements were extended to cover this aspect.

Since 1970 measurements of nutrient concentrations have been a routine part of oceanographic investigations off Natal, carried out initially by the National Physical Research Laboratory (NPRL), and then later by the National Research Institute for Oceanology (NRIO). This data set has been supplemented by measurements made during marine pollution monitoring investigations carried out by the National Institute for Water Research (NIWR). However, despite the large amount of data collected and the excellent supporting physical oceanographic information, no detailed analyses on the distribution of the chemical nutrients have been made. Pearce (1973a, b, 1977a) and Carter (1973a,b, 1977) have discussed aspects of the data but mainly in relation to physical and/or biological oceanographic studies. Oliff

(1973) has presented the most detailed study but as this was limited to the Richards Bay area the conclusions of the analysis cannot be applied to the whole Natal coastal ocean.

In terms of physical oceanographic coverage involving shipborne measurements three areas off the Natal coast have been intensively studied. These areas are centred at Richards Bay, Durban and Port Edward and are depicted in Figure 6.1. Schumann (Chapter 5) has discussed the physical oceanography of these areas in terms of topography and the effects of the Agulhas Current. The geographic spacing of the regions and the differing hydrographic regimes make these areas a convenient basis for comparative studies of the distribution, and the factors affecting the distribution, of chemical

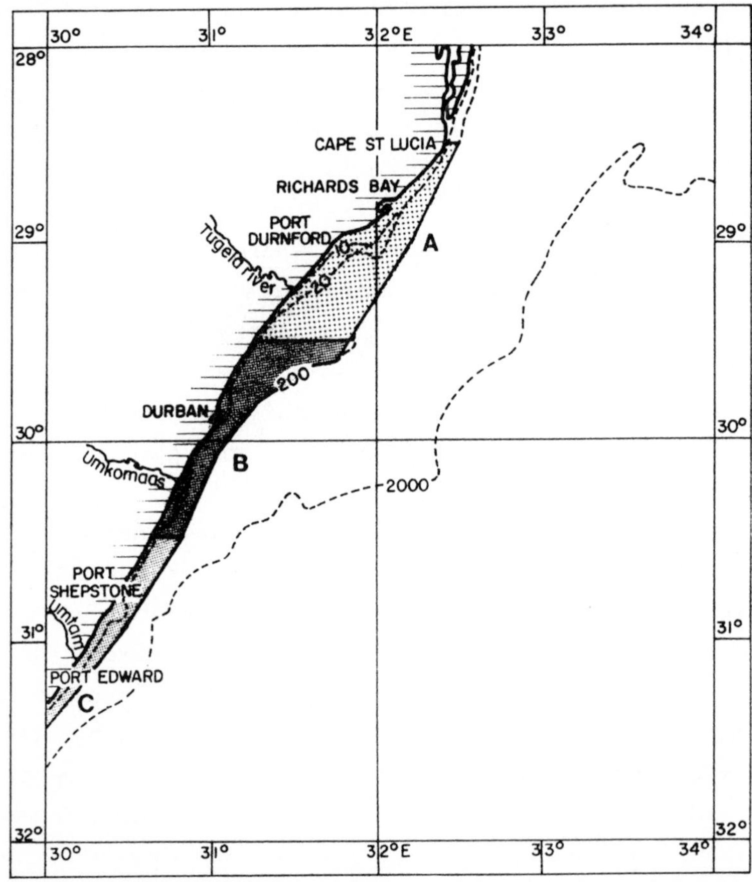

Figure 6.1 The three areas for which nutrient data were obtained, and which are discussed separately in the text, are shown, namely A: Richards Bay, B: Durban and C: Port Edward. Degree squares were used, with the restriction that water depth is less than 200 m.

nutrients off Natal; the concentration of the measurements themselves is also relevant here. This chapter concentrates on these aspects.

DATA

The data for the analysis presented here were obtained from the South African Data Centre for Oceanography (SADCO) operated by the NRIO. For ease of retrieval the SADCO data were accessed by one-degree squares centred at each of the three areas of interest (Figure 6.1). The offshore limit of the continental shelf is taken to be the 200 m isobath and the specifications of the data access were such that only data from stations where the water depth was 200 m or less were retrieved. The data discussed here, therefore, include all the data stored in SADCO for these areas, and not only the data from the NRIO or NIWR cruises, although these Institutes provided the bulk of the data. Both Institutes used similar sampling apparatus, namely a rosette sampler or 5 ℓ NIO bottles, with all the analyses being carried out by the NIWR on an AC 60 automated analysis system (Turner and Stanton, 1983). Table 6.1 lists the precision of the methods used in the analyses.

Table 6.1 Precision of the analytical methods used by the NIWR for the determination of nitrate-nitrogen, reactive silicate and phosphate-phosphorus (after Turner and Stanton, 1983). N represents the number of measurements.

Nutrient	Concentration µM/ℓ	Coefficient of Variation %	N
Nitrates	7.14	5.6	66
Silicates	5.00	1.8	48
	10.00	0.3	18
Phosphates	3.00	7.3	66

MEAN NUTRIENT CONCENTRATIONS AND VARIATIONS

Richards Bay

The degree square centred at Richards Bay from which the measurements for this analysis were taken is situated towards the northern boundary

Table 6.2 Maximum and minimum values, means and standard deviations of the concentrations of the dissolved inorganic chemical nutrients nitrate-nitrogen, reactive silicate and phosphate-phosphorus on the continental shelf off Richards Bay.

		Units µM/ℓ	
	Nitrates	Silicates	Phosphates
Maximum	19.64	13.21	8.07
Minimum	0.00	0.43	0.14
Mean	2.70	3.48	0.79
Standard deviation	3.04	2.36	0.67
Number of samples	2291	2297	2294

of the central shelf region as defined by Schumann (Chapter 5). The area is characterized as a wide shelf region relative to the generally narrow continental shelf of the east coast of South Africa. Table 6.2 lists some statistics of the chemical parameters measured in this area. The large ranges in nutrient concentration found, as well as the high variances indicated by the large standard deviations, show that nutrient distributions in the Richards Bay area are not uniform; this can be attributed to the vertical and horizontal gradients in nutrient concentrations found in the area.

As is generally the case, and as is illustrated by Oliff (1973), gradients in nutrient concentrations with depth off Richards Bay are positive. Table 6.3 shows this in the form of mean concentrations of nutrients for three depths in continental shelf waters off Richards Bay; values measured in the Agulhas Current are included for comparison purposes.

Table 6.3 Mean concentrations of inorganic chemical nutrients at three depths on the continental shelf and in the Agulhas Current off Richards Bay (modified from Oliff, 1973)

Depth		Units µM/ℓ	
(m)	Nitrates	Silicates	Phosphates
Continental Shelf			
10	1.84	3.20	0.94
50	3.66	4.40	1.05
100	7.58	7.00	1.35
Agulhas Current			
10	1.04	1.95	0.72
50	1.66	2.29	0.90
100	1.74	2.73	1.05

It can be seen from the table that nitrate-nitrogen and reactive silicate concentration gradients are stronger than that for phosphate-phosphorus. Phosphate-phosphorus, apart from being biologically labile, does not reach high concentrations in the upper layers of the ocean (e.g. Wyrtki, 1971); this lack of a strong gradient and the variability due to biological processes means that phosphate concentrations are unsuitable for demonstrating processes affecting the distribution of nutrients off Natal. Thus, although some statistics regarding the distribution of phosphate-phosphorus are presented here, the discussion concentrates on nitrate-nitrogen and reactive silicate.

Oliff (1973) and Pearce (1977a) demonstrated a negative gradient in nutrient concentrations with distance offshore. This is illustrated in Table 6.4.

Table 6.4 Mean concentration (μM/ℓ) of the inorganic chemical nutrients at 10 m depth at different distances offshore on the continental shelf off Richards Bay and in the Agulhas Current (modified from Oliff, 1973).

Nutrient	Inner-shelf	Mid-shelf	Outer-shelf	Agulhas Current
Nitrates	2.70	2.28	2.07	1.04
Silicates	4.38	3.66	3.78	1.95
Phosphates	1.10	1.01	0.98	0.72

These horizontal and vertical gradients are attributed to the actual origin of the water on the continental shelf, sporadic upwelling that occurs up against the continental margin, and biological modification of the nutrient levels. Pearce (1973a) showed, according to temperature/salinity values, that Natal continental shelf waters are derived from mid-depth in the Agulhas Current. In such a western boundary current in the southern hemisphere colder and denser waters at the same depth lie to the right in the direction of flow; the effects of this are evident in temperature sections across the Agulhas Current (e.g. Figure 5.5). Because of the positive gradient in nutrient concentration with depth, inshore surface nutrient concentrations are therefore higher than surface values offshore in the Agulhas Current (Table 6.4).

The horizontal gradients observed on the continental shelf itself are probably due to this feature combined with meandering of the Agulhas

Current and upwelling. With regard to the meandering, Pearce (1977a) showed that although generally situated near the 200 m isobath, the western boundary of the Agulhas Current has been observed inshore of the 100 m isobath on a number of occasions. This would have led to Agulhas Current water, with its low nutrient concentrations (Table 6.4; Wyrtki, 1971), being located on the outer shelf, thus leading to the observed negative horizontal gradient with distance offshore in nutrient concentrations.

Pearce (1977a) and Oliff (1973) have reported sporadic local upwelling off Richards Bay, with temperature changes at the surface of about 1°C. Pearce (1977a) calculated, on the basis of 86 days of observations spread throughout the calender year, that such upwelling or its effects could be observed in the area for 31 per cent of the time. The upwelling effects are generally restricted to the nearshore area, with the areal extent of the upwelling rarely exceeding 200 km^2 (Oliff, 1973; Pearce, 1977a). Further, the duration of upwelling events appears to be generally of the order of one day, although the persistence of the effects may last longer, depending upon the currents prevailing at the time. In these upwelling events water from around 100 m depth is drawn to the surface (Pearce, 1977a). This can lead to increases in surface concentrations of nitrate-nitrogen from 1.0 to 7.0 $\mu M/\ell$ (Oliff, 1973) although the levels usually reached are of the order of 4.0 $\mu M/\ell$ (Pearce, 1977a). Pearce (1973b, 1977a) could not find a consistent relationship between upwelling and winds, shelf currents, Agulhas Current meandering or other factors in the Richards Bay area. This is probably due to inadequacies in the data set rather than to the lack of such relationships.

Salinities of nearshore surface waters off Richards Bay may be depressed during summer by land run-off through the Richards Bay estuary (Pearce, 1977a). Salinities can be reduced by up to two parts per thousand which represents a dilution of approximately 5 per cent. Peak nitrate-nitrogen levels within Richards Bay estuary attain 20 $\mu M/\ell$ (Begg, 1978) and, taking the dilution factor into account, a maximum increase of around 1 $\mu M/\ell$ can therefore be expected in nearshore waters during periods of strong outflow. This value is small relative to the scale of variability and is invisible both in the nutrient values and in the individual values for the nearshore waters presented by Oliff

(1973).

These upwelling events, the current regime and biological activity in the area all contribute, on both temporal and spatial scales, to patchiness in the nutrient distributions. As a consequence, plankton production rates and biomass distribuitions in the region are also patchy. This is discussed further in Chapter 7.

Durban

The degree-square used for data extraction is centred on Durban (Figure 6.1) and incorporates the Durban shelf as defined by Schumann (Chapter 5). Although, by local standards of investigation, the ocean adjacent to Durban has received extensive attention, routine chemical nutrient measurements were instituted in the area only in 1973. Further, the post-1973 oceanographic surveys were generally aimed at investigations of the Agulhas Current (e.g. Pearce, 1977b) which is usually situated more than 40 km offshore and thus the continental shelf here has not received as extensive coverage as has the Richards Bay area.

Table 6.5 lists the relevant statistics on the inorganic chemical nutrient concentrations on the continental shelf off Durban. As is the case off Richards Bay, there is a large variation in the concentrations, this being due to strong vertical gradients found in the region. Examples of these gradients are shown in Figure 6.2. It is also evident from the figure that there are quite marked day-to-day variations in the profiles, which are due either to upliftment or sinking of the region of maximum gradient in the profile or upliftment plus enrichment of the surface layers. The former feature is shown particularly well in the profiles for March, 1974. Schumann (Chapter 5) has presented evidence that water-column dynamics in the Durban shelf region are driven by perturbations in the Agulhas Current rather than by local wind events. The dominant effect of these perturbations is the presence of a semi-permanent cyclonic eddy situated 20 to 40 km off Durban (Pearce, *et al.* 1979). This eddy generally has colder water at its centre, due to upwelling, and is thus observable in satellite imagery of the area (e.g. Figure 5.10). Figure 6.3 shows vertical profiles of nitrate-nitrogen and temperature measured in the preferred

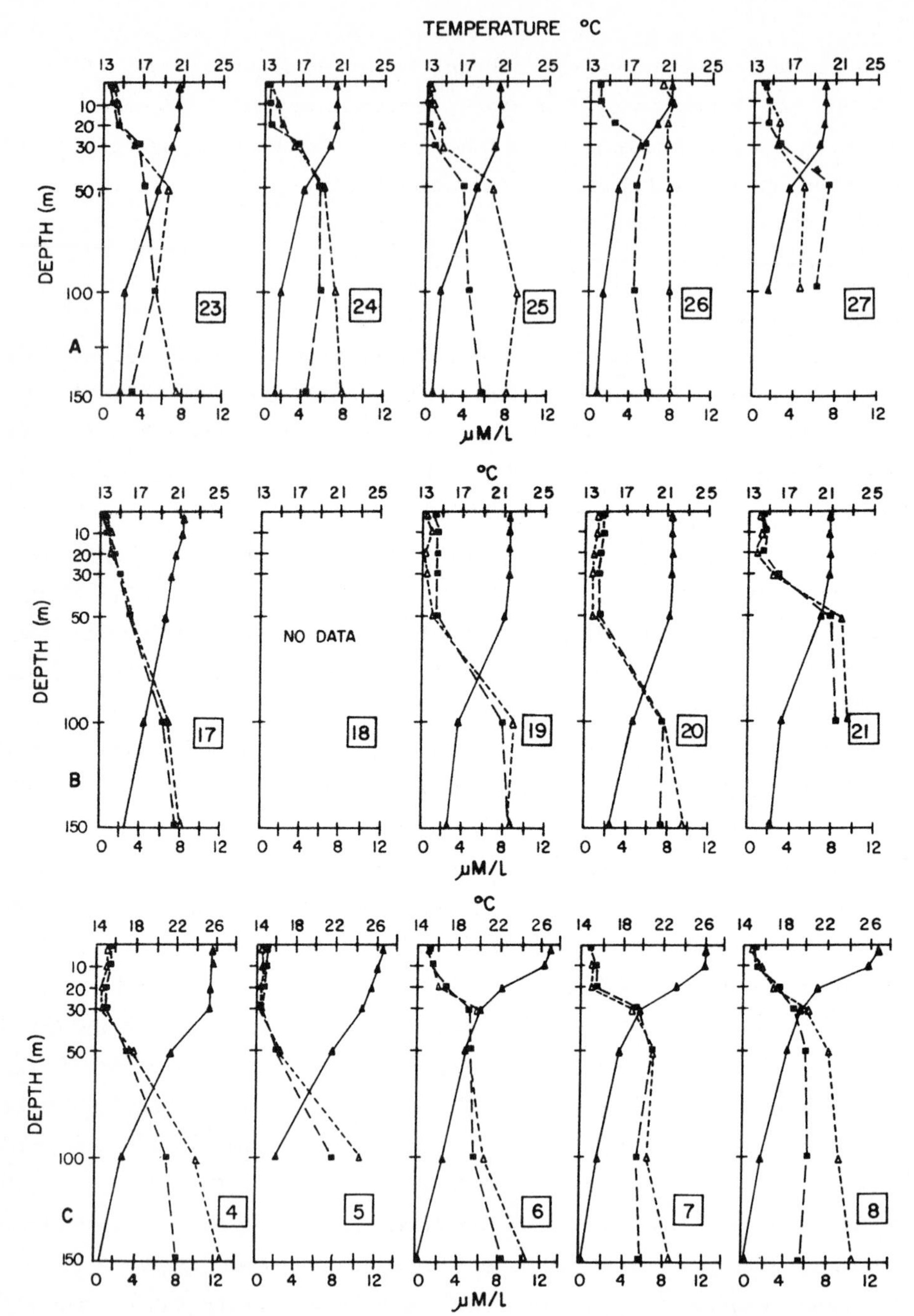

Figure 6.2 Five-day time series of vertical profiles of nitrate-nitrogen (△-----△), reactive silicate (■----■) and temperature (▲——▲) at a station location 10 km offshore from Durban in 180 m water depth for cruises carried out in July (A) and September (B), 1973, and March, 1974 (C) (data from Pearce, 1976).

Table 6.5 Maximum and minimum values, means and standard deviations of the concentrations of the dissolved inorganic chemical nutrients nitrate-nitrogen, reactive silicate and phosphate-phosphorus on the continental shelf off Durban.

	Units µM/ℓ		
	Nitrates	Silicates	Phosphates
Maximum	16.79	16.71	4.20
Minimum	0.07	0.03	0.00
Mean	3.33	3.71	0.62
Standard deviation	4.36	2.76	0.53
Number of samples	664	668	673

location of this eddy in September, 1973. On the 18th September there was no eddy present in the area and the temperature profiles show well-developed surface mixed layers extending to about 50 m, with the maximum gradient in the nitrate-nitrogen profiles lying below 60 and 100 m. On the 19th September an eddy was detected (Pearce, 1974) and the mixed layer depths were reduced to around 30 m at both stations. Surface temperature also showed a decline of the order of 1°C. More dramatically, however, the depth of maximum gradient in the nitrate-nitrogen profiles had been lifted to about 40 m at both stations, which represents a rise of 60 m at the 40 km station. On the 20th September the eddy had dissipated or moved out of the area with the result that the mixed layer had deepened and the nitrate-nitrogen profiles displayed slacker and deeper gradients.

The eddy discussed above had marked manifestations as regards surface temperature and appeared to be 20 km in diameter (Pearce, 1974). Smaller eddies (3-10 km diameter) than this have been observed in the area, three of which were apparent between 10 and 25 km offshore on the 7th March, 1974 (Pearce, 1974). From the above it is apparent that the eddies observed were probably the cause of the decrease in the depth of the mixed layer and of the associated upliftment of the nitrate-nitrogen gradients demonstrated for the 5th to 8th March in Figure 6.2. The frequency of occurence of these smaller eddies is unknown.

As pointed out above, most of the research cruises on which nutrients were measured were directed at studying the Agulhas Current or its western boundary. As the Agulhas Current is generally situated some distance offshore, smaller-scale horizontal gradients in chemical concentrations extending from nearshore to the 200 m isobath have not

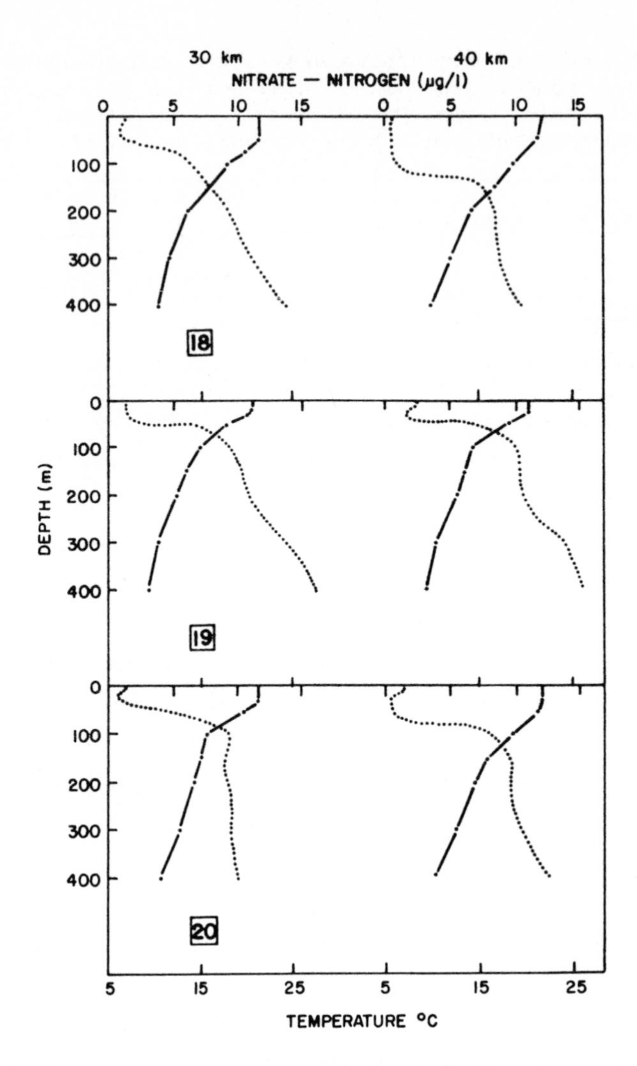

Figure 6.3 Changes in vertical profiles of nitrate-nitrogen (dashed line) and temperature (solid line) associated with the formation of a cyclonic eddy in September, 1973, at stations 30 and 40 km offshore from Durban (data from Pearce, 1976).

been measured (cf. Richards Bay). However, Burchall (1968) measured nitrate-nitrogen concentrations in the water column at a depth of 50 m, 7 km off Durban on 11 occasions in the period August 1965 to June 1966 and found a mean concentration of 2.32 (sd = 2.64) μM/ℓ. In comparison the mean nitrate-nitrogen concentration in the 0 to 50 m depth for the Durban shelf waters within the degree-square discussed here was 2.22 (sd = 2.43) μM/ℓ and as most of measurements (300 out of a total of 424) were made at a station located 10 km offshore in 180 m depth (Pearce, 1976), it appears that horizontal gradients in the region are generally weak. This accords with the apparent absence of an upwelling site against the coast (cf. Richards Bay).

Carter (1977) showed that the seasonal variation in mean surface temperature off Durban ranged from a low of 21°C in late winter to 26°C in summer. Surface salinity profiles along a 100 km long transect off Durban also indicate that lower salinities are associated with the high summer temperatures (Pearce, 1974). These effects extend across the Agulhas Current and thus appear to be due to variations in the source water of the current rather than to local fresh water inflow and therefore the entire region would be affected. Possible covariation in nutrient concentrations is discussed in the section on seasonal variation below.

Port Edward

The degree-square from which data were extracted from SADCO for analysis encompasses the southern shelf as defined by Schumann (Chapter 5), and being comparatively remote from major urban areas or ports this area has not received extensive scientific attention. However, data have been collected from a 60-km-long transect off Port Edward by NRIO (Figure 5.1), nine cruises being made to this area in the period April 1974 to April 1975 and a total of 19 crossings of the transect achieved (Lundie, 1976). Three station locations on the inshore end of the transect were on the continental shelf and these supplied the bulk of the data discussed here. These data are supplemented by measurements made during NIWR marine pollution monitoring cruises to the area.

Table 6.6 lists some statistics describing the chemical concentrations measured in the area.

Again the variability in all the concentrations is high and, as with the two other areas discussed, this is attributable to strong vertical gradients in the chemical distributions. This is illustrated in Figure 6.4 which presents examples of nitrate-nitrogen, reactive silicate and temperature profiles measured off Port Edward. It is evident in these profiles that the degree of development of the upper mixed layer is one

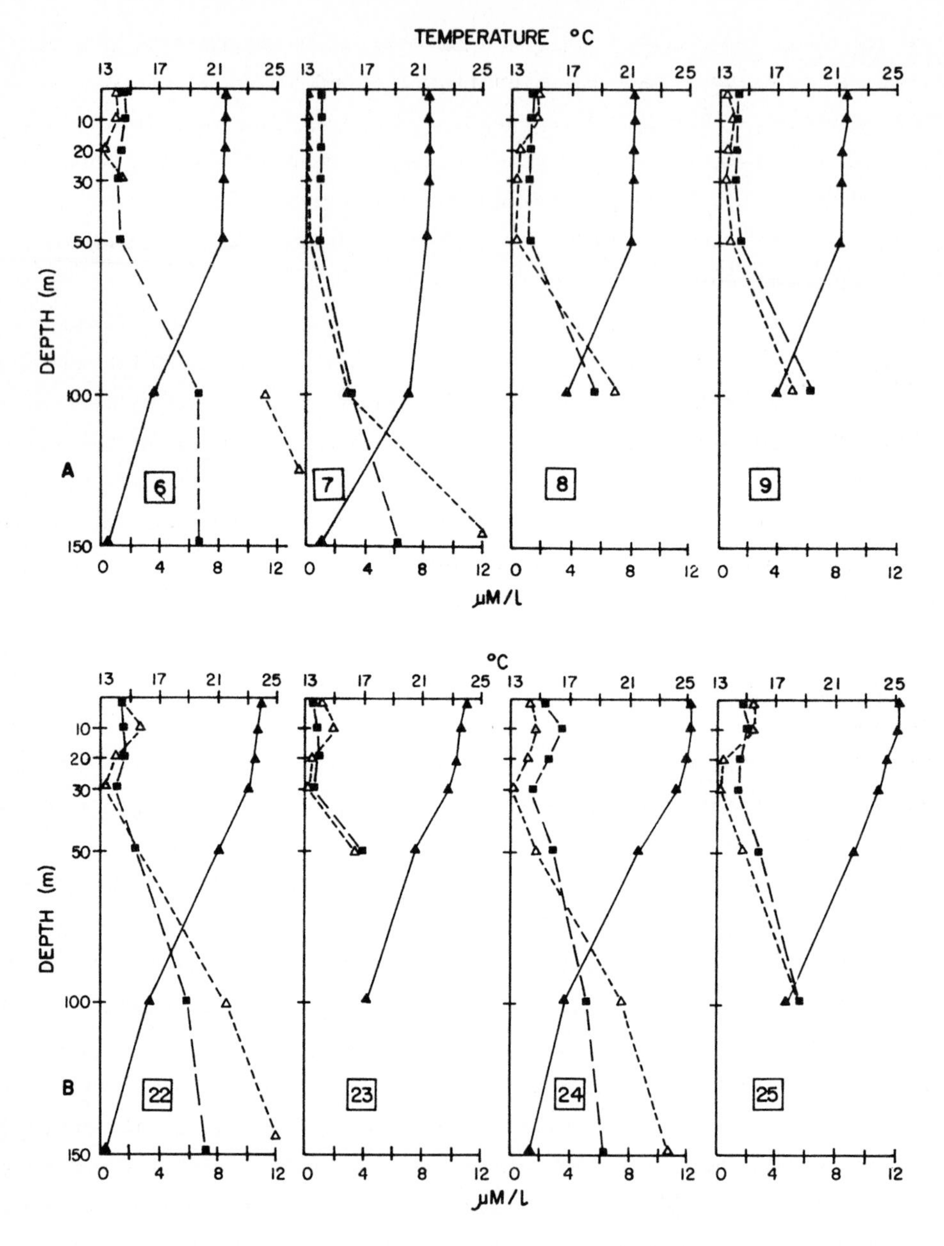

Figure 6.4 Four-day time series of vertical profiles of nitrate-nitrogen (Δ- - - - -Δ), reactive silicate (■— — —■) and temperature (▲———▲) at a station location 10 km offshore from Port Edward for cruises carried out in August (A), 1974 and April (B), 1975 (data from Lundie, 1976).

Table 6.6 Maximum and minimum values, means and standard deviations of the concentrations of the dissolved inorganic chemical nutrients nitrate-nitrogen, reactive silicate and phosphate-phosphorus on the continental shelf off Port Edward.

	Nitrates	Units µM/ℓ Silicates	Phosphates
Maximum	15.80	8.65	5.80
Minimum	0.10	0.06	0.00
Mean	2.48	2.98	0.61
Standard deviation	2.56	2.08	0.46
Number of samples	593	593	592

factor which affects the profiles and that cold bottom-water intrusion is another.

Carter (1977) and Schumann (Chapter 5) have shown that the western boundary of the Agulhas Current off Port Edward frequently lies over the continental shelf. The station location where the profiles were measured was within 2 km of the shelf break and the depth of the mixed layer together with high surface temperatures shown in the profiles indicate intrusion by the Agulhas Current. Low nutrient concentrations are generally found in the surface mixed layer (e.g. profiles for 6th to 9th August) and thus the shapes of the profiles are affected by the depths of the intrusions.

In the profiles displayed for the 22nd to 25th April, 1975, the upper 30 m show weak gradients in temperature, as opposed to isothermal conditions, and nutrient minima at the 20 to 30 m depth range. These profiles persist throughout the series despite an approximately 1°C warming of the upper 50 m between the 23rd and 24th April. Current speeds on all four days ranged from 1.3 to 1.95 m/s and flows were uniformly south-ward (Lundie, 1976). In the light of this, and the turbulence expected from such high currents, the gradients observed are perplexing and no explanation is offered. The feature is a further indication of the variabilities of nutrient concentrations in the area.

The vertical profiles shown in Figure 6.5 indicate that high nutrient concentrations can occur at the bottom of the water column. These high values are usually associated with low temperatures (e.g. profiles for the 7.8 and 22.4). Schumann (1986) has shown that this is due to Ekman veering in the bottom boundary layer bringing colder water onto the shelf.

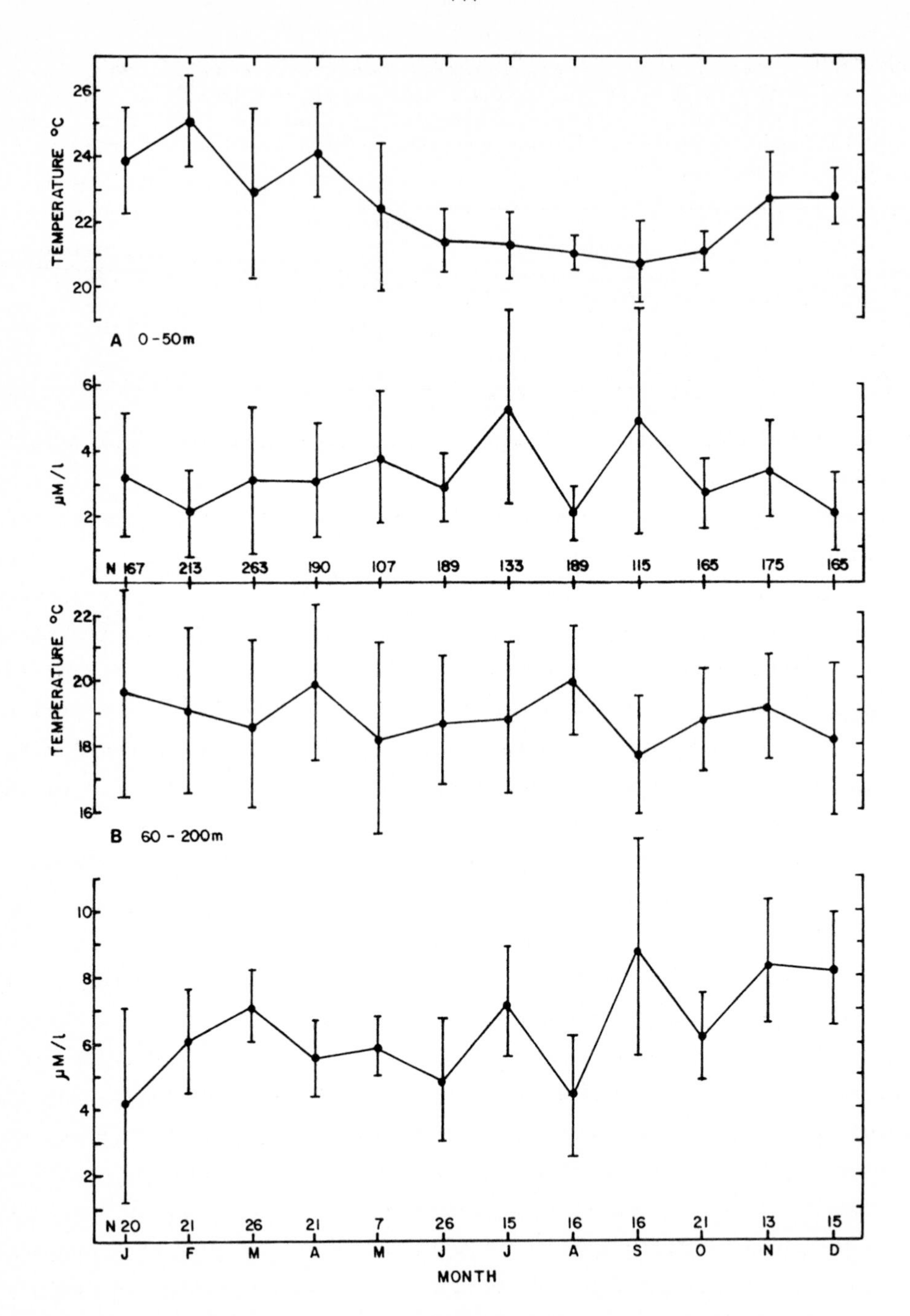

Figure 6.5 Monthly means of temperatures and reactive silicate in the 0-50 m (A) and 60-200 m (B) depth ranges on the continental shelf off Richards Bay (data from SADCO). N is the number of measurements in each month.

This cold water does upwell against the coast, as is evident in Figure 5.7 and also shown by Carter (1977). This upwelling leads to the development of negative nutrient concentration gradients across the continental shelf even though the distance involved is short. Examples are presented in Table 6.7.

Table 6.7 Mean nitrate-nitrogen and reactive silicate concentrations in the surface layers (maximum depth = 50 m) at locations 2, 5 and 10 km offshore on the continental shelf off Port Edward (data taken from Lundie, 1976).

Cruise dates	Stn location (km offshore)	Units µM/ℓ Nitrates	Silicates	Number of Observations
1974				
May	2	1.92	3.10	6
	5	1.27	2.64	12
	10	1.73	2.87	12
August	2	4.19	3.87	8
	5	1.76	2.54	16
	10	0.70	1.29	16
1975				
February	2	1.80	2.33	8
	5	1.57	1.94	16
	10	1.88	2.18	16
April	2	4.88	4.19	8
	5	3.55	3.21	15
	10	1.50	2.02	16

These gradients are not strong in every case and are obviously subject to temporal variation. The area, however, does exhibit stability on time scales of days. This is illustrated in Table 6.8 which shows the variation in horizontal gradients of nitrate-nitrogen across the continental shelf over four days in August, 1974. The gradients were more or less stable over the entire period; the moored current-meter results for this area presented in Figure 5.12 also reflect a steady situation.

Although there is the expected seasonal variation in surface temperature, no coherent seasonal variation in salinities is apparent in the NRIO data for the continental shelf off Port Edward (Lundie, 1976; Carter, 1977). Any variation that may be present is masked by local upwelling effects or intrusions of Agulhas Current surface water across the shelf.

Table 6.8 Mean nitrate-nitrogen concentrations (μM/ℓ) over the 0-50 m depth range for three stations on a transect across the continental shelf off Port Edward for the period 6-9 August, 1974 (data taken from Lundie, 1976).

Station location (km offshore)	Date			
	6 August	7 August	8 August	9 August
2	3.2	3.9	5.9	3.8
5	1.5	1.8	1.9	1.9
10	0.9	0.3	0.8	0.8

COMPARISON BETWEEN THE REGIONS

All three regions displayed wide variations in concentrations and distributions of the inorganic chemical nutrients. It was shown that different processes contributed to this variability, e.g. coastal upwelling off Richards Bay and Port Edward, eddy-centre upwelling off Durban, and varying degrees of intrusion of Agulhas Current surface water. Student's t test (Zar, 1974) was applied to the means and standard deviations of the nutrient concentrations presented in Tables 6.2, 6.5 and 6.6 to determine whether there were any differences in the magnitude of the means between the areas. The results are presented in Table 6.9.

Table 6.9 Hierarchical comparison of mean inorganic nutrient concentrations, over the 50-0 m and 200-0 m depth ranges, measured on the continental shelf off Richards Bay (RB), Durban (DBN) and Port Edward (PtE). Directional magnitude signs (>) show that the means are different at the 95% confidence interval and in which region the higher mean occurs.

Nutrient	Depth range	
	50-0 m	200-0 m
Nitrates	RB=DBN > Pte	DBN > RB > PtE
Silicates	RB=DBN > PtE	DBN > RB > PtE
Phosphates	RB > DBN, RB > PtE, DBN > PtE	RB > DBN=Pte

The Table shows that, in the 50-0 m depth range, the wide continental shelf regions off Richards Bay and Durban are characterized by higher mean nutrient levels than is the narrow shelf area off Port Edward. These differences are attributed to the greater influence of Agulhas Current surface water in the latter region.

These differences between the shelf regions are also evident in the

200-0 m depth range where the Port Edward region is again characterized by lower mean levels of all three nutrient species. However, whereas the Durban and Richards Bay mean nitrate-nitrogen and reactive silicate concentration were similar for the 50-0 m depth range, these values were higher off Durban in the 200-0 m range. These differences are also reflected in the 100 to 200 m depth range. The high values off Durban, when considered together with the positive gradients in nutrient concentrations with increase in depth found off Natal (e.g. Figure 6.1), indicate that more colder water is brought up onto the shelf off Durban than off Richards Bay. It has been shown that the semi-permanent eddy located east of Durban (Pearce et al. 1979) can cause significant uplift of colder water at its centre. It is apparently this mechanism that leads to the higher nutrient levels off Durban.

A further difference between the areas is that the cross-shelf gradients in nutrient concentrations are well developed off Richards Bay and Port Edward, but appear to be weak off Durban. The well-developed gradients are due to upwelling against the coastal margin as opposed to the eddy-centre upwelling off Durban. Schumann (1986 and Chapter 5) has related the coastal upwelling off Port Edward to an onshore Ekman veering in the bottom boundary layer associated with a strong southward flow. Pearce (1973b; 1977) found the coastal upwelling off Richards Bay to be sporadic and could not show a strong causal relationship with any of the features measured, although here again strong southward flow was a contributor. The three areas discussed therefore have different mechanisms affecting the nutrient levels as well as differences in the mean nutrient concentrations. However, the region as a whole is characterized by short-term temporal variations in nutrient levels, in common with the variability displayed for the physical variables.

SEASONAL VARIATION

The distributions of the chemical nutrients on the Natal continental shelf have been shown to vary in both time and space. To investigate whether there is any seasonal signal that overrides the short-term variability, the distribution of monthly mean concentrations of reactive silicate and temperature in two depth ranges off Richards Bay are compared in Figure 6.5. The Richards Bay area was selected because of the larger number of measurements made there and silicate concentration was used because of the gradients observed and because it is the most conservative of the three nutrients investigated.

Figure 6.5a shows that although there are summer maxima and late winter minima in temperature in the upper 50 m in the water column, there is no corresponding seasonality in the silicate concentrations. Further, Figure 6.5b shows that below 50 m the seasonal temperature signal has disappeared and, again, there is no seasonality in the silicate distribution. The number of nutrient measurements made in the 60 to 200 m depth range for any one month is low, however, and, in fact, represents the coverage obtained in one of the NRIO monthly cruises to the area (e.g. Pearce, 1973b). The monthly mean concentrations displayed for this depth range may thus be affected by one or more episodic events, such as storms (e.g. McLaughlin *et al.* 1975). Therefore, although no seasonality is apparent such longer-period cyclicity cannot be ruled out on the basis of the data presented here for the Natal continental shelf.

Despite these drawbacks the data presented in Figure 6.5 indicate that nutrient distributions on the Natal continental shelf are dominated by event scale rather than longer-term cyclical processes. This is essentially the conclusion reached by Pearce, *et al.* (1979) and Pearce and Grundlingh (1982) from their analyses of currents off the Natal coast. In this aspect the region differs from its eastern North American counterpart, the North Carolina shelf. Here Atkinson, *et al.* (1982, 1984) have shown that the waters inshore of the western boundary of the Gulf Stream are characterized by similar ranges of nutrients but that the nutrient levels are affected by summer upwelling and river outflow in winter. Moreover, short-term variability does not seem to be as severe as off Natal.

In these respects Natal continental shelf waters also differ from those of the south western Cape, South Africa, in that the latter are strongly dominated by seasonal upwelling (Andrews and Hutchings, 1980). A further difference is the high concentration of nutrients found at the surface during upwelling events (15 μM NO_3-N/ℓ) (e.g. Barlow,

1984). However, coupled with these high levels are rapid modification by phytoplankton growth and wind-induced mixing of the surface layers (Andrews and Hutchings, 1980). Thus, although dissimilar in seasonal distributions and maximum concentrations found in the surface layers, the waters in both regions show strong variability on short temporal scales.

CONCLUSIONS

The three areas of the Natal continental shelf are characterized by spatial and temporal variability in the distributions of the chemical nutrients. The scales of the variability are small, being of the order of tens of kilometres and days, respectively. Although the ranges of nutrient concentrations are similar in all three areas, which is a consequence of the origin of the waters on the continental shelf the mean nutrient concentrations on the wider shelf regions off Richards Bay and Durban are higher than those on the narrow shelf off Port Edward. This is attributed to the greater influence of the Agulhas Current in the latter area. The Durban shelf region also has higher mean nutrient concentrations below 50 m depth than does the Richards Bay area, and this is attributed to the presence of a semi-permanent cyclonic eddy situated offshore of the 200 m isobath which draws deep, nutrient-rich water up onto the shelf off Durban. The Natal continental shelf exhibits no seasonal nutrient variation and land runoff effects were not apparent in the nutrient concentrations measured. The most important feature from a plankton production point of view is the extreme variability.

REFERENCES

ANDREWS, W R H and L HUTCHINGS (1980). Upwelling in the southern Benguela Current. **Progress in Oceanography,** 19, 1-81.

ATKINSON, L P, L J PIETRAFESA and E H HOFFMAN (1982). An evaluation of nutrient sources to Onslow Bay, North Carolina. **Journal of Marine Research,** 40, 679-699.

ATKINSON, L P, P G O'MALLEY, J A YODER and G-A PAFFENHOFER (1984).The effect of summertime shelf break upwelling on nutrient flux in south eastern United States continental shelf waters. **Journal of Marine Research,** 42, 969-993.

BARLOW, R G (1984). Dynamics of the decline of a phytoplankton bloom after an upwelling event. **Marine Ecology Progress Series,** 16, 121-126.

BEGG, G (1978). **The estuaries of Natal.** Natal Town and Regional Planning, vol. 41, 657 pp.

BURCHALL, J (1968). An evaluation of primary productivity studies in the continental shelf region of the Agulhas Current near Durban (1961-1966). **Oceanographic Research Institute Investigational Report,** 21, 44pp.

CARTER, R A (1973a). Plankton studies at Richards Bay. **CSIR NPRL Oceanography Division Contract Report No. C FIS 378,** Durban, South Africa, 13 pp.

CARTER, R A (1973b). Factors affecting the development and distribution of marine plankton in the vicinity of Richards Bay. South African National Oceanography Symposium, Cape Town. **CSIR NPRL Oceanography Division Contribution No. 30,** 34 pp.

CARTER, R A (1977). **The distribution of calanoid copepoda in the Agulhas Current system off Natal, South Africa.** MSc thesis, University of Natal, 165 pp.

LUNDIE, G S H (1976). Data report for the series of nine cruises in project 'Ported' off Port Edward 1974 to 1975. **CSIR NRIO Internal General Report SEA IR 7609.**

McLAUGLIN, D, J A ELDER, G T ORLOB, D F KIBLER and D E EVENSON (1975). A conceptual representation of the New York Bight ecosystem. **NOAA Technical memorandum ERL MESA-4 PB-252 43.**

OLIFF, W (1973). Chemistry and productivity at Richards Bay. **CSIR NPRL Oceanography Division Contract Report No. C FIS 37B,** Durban, South Africa, 17 pp.

PEARCE, A F (1973a). Water properties and currents off Richards Bay. South African National Oceanographic Symposium. Cape Town. **CSIR NPRL Oceanography Division Contribution No. 29.** 28pp.

PEARCE, A F comp (1973b). Technical report on special current studies at Richards Bay. **CSIR NPRL Oceanography Division Contract Report No. C FIS 37B,** Durban, South Africa, 424 pp.

PEARCE, A F (1974). Preliminary results from a set of ten cruises out to 100 km off Durban. Part 1: Surface temperature, salinity and currents. CSIR NRIO Internal General Report IG 74/11. 39 pp.

PEARCE, A F (1976). Data report for the series of ten cruises in project 'Interface' off Durban 1972 to 1974. **CSIR NRIO Internal**

General Report SEA IR 7606.

PEARCE, A F (1977a). **The shelf circulation off the east coast of South Africa.** MSc thesis, University of Natal, 220 pp.

PEARCE, A F (1977b). Some features of the upper 500 m of the Agulhas Current. **Journal of Marine Research,** 35, 731-753.

PEARCE, A F, E H SCHUMANN and G S H LUNDIE (1979). Features of the shelf circulation off the Natal coast. **South African Journal of Science,** 74, 328-331.

PEARCE, A F and M L GRÜNDLINGH (1982). Is there a seasonal variation in the Agulhas Current? **Journal of Marine Research,** 40, 177-184.

SCHUMANN, E H (1986). Bottom boundary layer observations inshore of the Agulhas Current. **South African Journal of Marine Science, 4, 93-102.**

TURNER, WD and R C STANTON (1983). Automatic analysis for marine nutrients based on the AC60 analyser. **NIWR Project Report 6701/6731,** 22pp.

WYRTKI, K (1971). **Oceanographic atlas of the International Indian Ocean expedition.** National Science Foundation, USA, 531 pp.

ZAR, J H (1974). **Biostatistical analysis.** Prentice-Hall International Inc., London, 620 pp.

Chapter 7

PLANKTON DISTRIBUTIONS IN NATAL COASTAL WATERS

Robin A Carter
National Research Institute for Oceanology, CSIR

Michael H Schleyer
Oceanographic Research Institute, Durban

INTRODUCTION

The plankton of Natal coastal waters has received little attention compared with the research effort devoted to physical oceanographic studies in the region (see Chapter 5). Further, the studies that have been made have generally been either part of large-scale surveys in the southwestern Indian Ocean (e.g. Zoutendyk, 1960, 1970; Burchall 1968a), or they have been single surveys, and have thus not taken account of the variability of the area (e.g. De Decker, 1964; Thorrington-Smith, 1969). Exceptions to this are the studies by Burchall (1968b) who investigated temporal variation in phytoplankton production off Durban, and Schleyer (1976, 1985) and Carter (1977) who researched temporal and spatial variations in zooplankton biomass and species distributions along transects across the Agulhas Current.

Subsequent plankton studies have concentrated upon the role of production in the trophic dynamics of a nearshore reef system (Schleyer, 1979, 1981) and distribution and dynamics in the region of a semi-permanent cyclonic eddy off Durban (NRIO, unpublished). Attention has also been focused on pollution indicators in the plankton (e.g. Gardner _et_ _al_. 1983) and on the distribution of specific zooplankton groups such as Pontellid copepoda (Connell, pers. comm.).

In this chapter major features of the published and unpublished results of plankton research in Natal continental shelf waters are reviewed and attempts are made to identify the major dynamic processes affecting plankton in the region.

PHYTOPLANKTON

Phytoplankton studies off Natal have been described by a number of researchers, and incorporate biomass, production and species distribution investigations. Figure 7.1 shows the locations of the studies reviewed here.

Biomass

There are few published estimates of phytoplankton biomass; chlorophyll-a concentrations have been given by Lundie (1979) and by Schleyer (1979, 1981) and phytoplankton cell counts by Schleyer (1981). Unpublished sources are more extensive, however, and these have been incorporated into Table 7.1 which lists chlorophyll-a concentrations recorded in Natal coastal waters.

Table 7.1 Means, standard deviations (SD) and ranges of chlorophyll-a measurements (all in µg/ℓ) made in Natal continental shelf waters. The location of each of the study sites is shown in Figure 7.1. (NRIO = National Research Institute for Oceanology, ORI = Oceanographic Research Institute). N is the number of samples.

Source	Location	Date	Depth (m)	Mean	SD	Range	N
NRIO (unpub)	Richards Bay	Jan. 1972	10	0.37	0.09	0.06-1.52	73
		Mar. 1972	10	0.25	0.08	0.03-0.94	100
	Tugela River	June 1978	0-15	0.85	0.12	0.52-1.04	44
	Cyclonic eddy off Durban	Mar. 1979	0-50	0.64	0.21	0.30-1.35	60
		Apr. 1979	0-50	0.64	0.32	0.18-1.83	99
		Sep. 1979	0-50	1.23	0.74	0.08-2.46	26
		Apr. 1980	0-50	0.53	0.20	0.25-0.91	21
		Aug. 1980	0-50	1.05	0.91	0.15-3.67	99
Lundie (1979)	Mkomasi River	Mar. 1979	0-50	0.64	0.20	0.30-1.14	51
Schleyer (1979,1981)	ORI Reef Durban	Mar. 1978-Apr. 1979	1	2.13	-	0.76-3.88	25

Table 7.1 shows that the chlorophyll-a concentrations measured are low with correspondingly small variation; the overall range for all 598 measurements is 0.03 to 3.88 µg/ℓ. Furthermore, although the spatial and temporal coverage of the measurements is large there are no significant differences between the mean chlorophyll concentrations listed in the table (Student's t test, Zar, 1974).

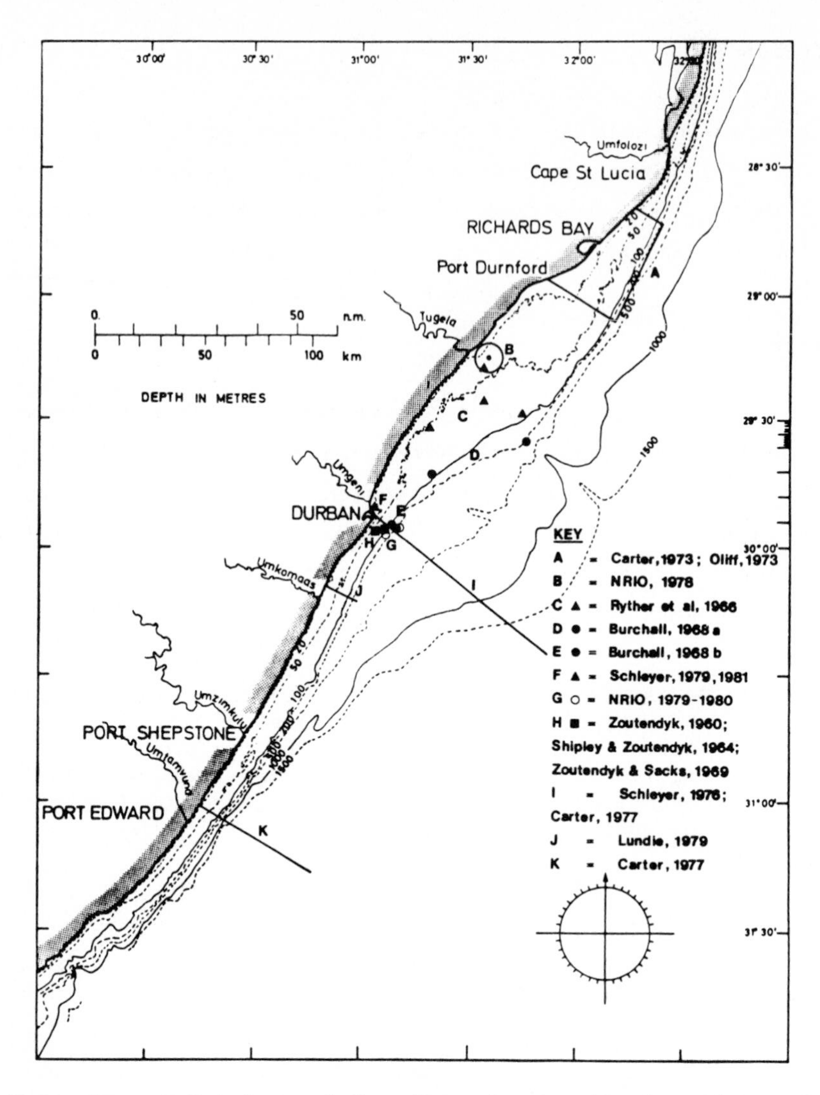

Figure 7.1 Map of Natal coastal waters showing the locations of the plankton investigations discussed in this chapter.

The small ranges of the NRIO measurements made in the region of the cyclonic eddy off Durban in 1979 and 1980 indicate weak gradients in chlorophyll concentrations with depth, which is borne out by the vertical profiles depicted in Figure 7.2. This figure shows that subsurface chlorophyll maxima, although present, are poorly developed, compared to distributions found on the Agulhas Bank, south of the subcontinent (Carter et al. 1986), and in continental shelf areas elsewhere (e.g. Holligan et al. 1984). The temperature gradients

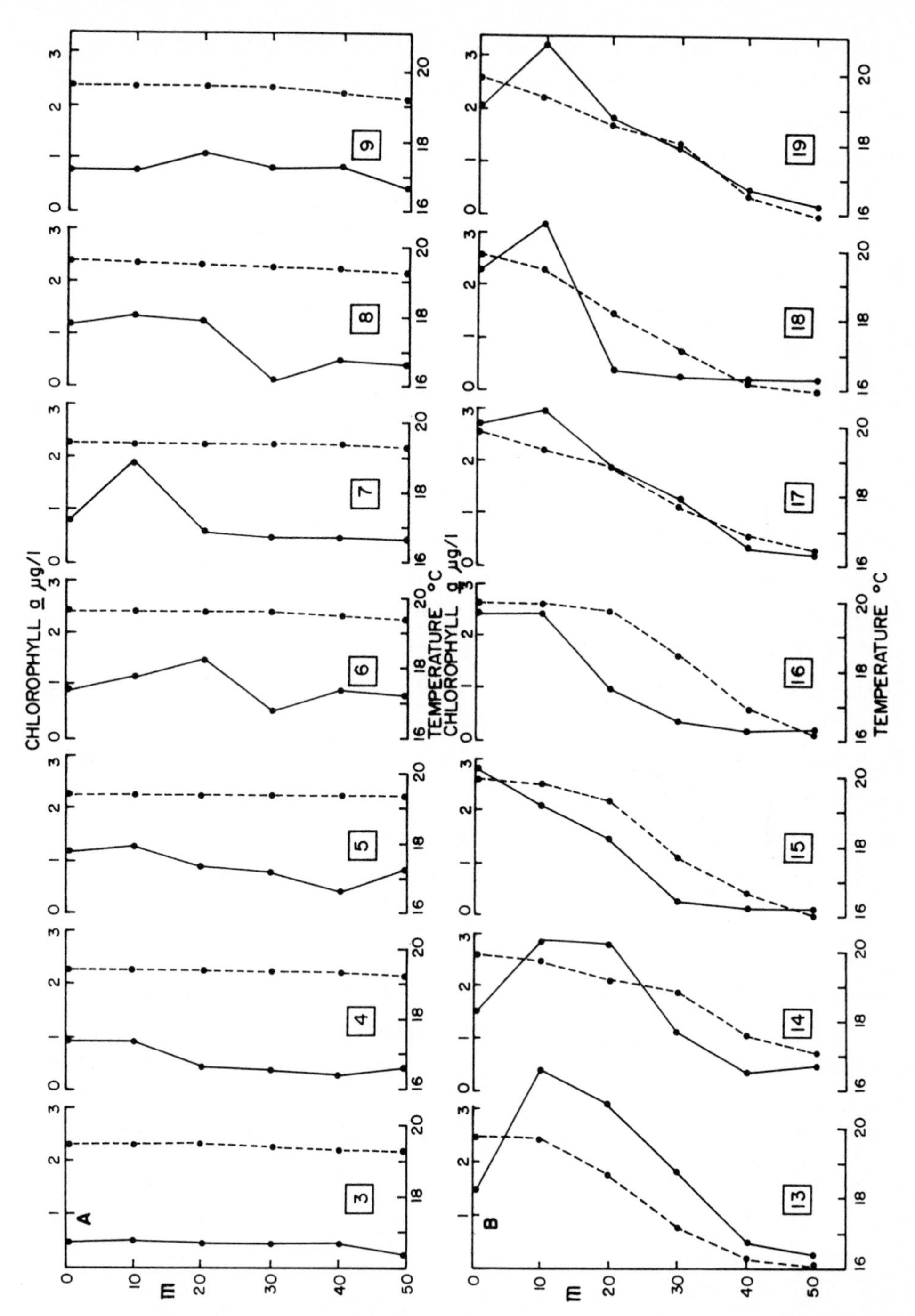

Figure 7.2 Chlorophyll-a (solid line) and temperature (dashed line) profiles measured in the vicinity of the semi-permanent cyclonic eddy off Durban. Part A refers to a cruise on 29 and 30 August, 1980, and b to 31 August to 1 September, 1980; station numbers are given in blocks.

displayed in Figure 7.2 are weak and therefore the chlorophyll maxima that do occur are probably the result of light inhibition of phytoplankton production at the surface.

The only other 'pure' phytoplankton biomass parameter reported for Natal waters are the cell counts, based on water as opposed to net samples, made by Schleyer (1981) for waters overlying the ORI reef and those made by NRIO (unpublished) in the vicinity of the cyclonic eddy off Durban (Figure 7.1) in September, 1979. Schleyer reported phytoplankton cell counts ranging from 1.5 to 5.4 x 10^6 cells/litre, the samples being dominated by microflagellates. In contrast, counts of 1 to 5 x 10^4 cells/litre were obtained in the vicinity of the cyclonic eddy, with diatoms dominating the samples. The two estimates are not entirely comparable, however, as Schleyer used AODC (Hobbie _et al_. 1977), whereas light microscopy on formalin-preserved samples was used to obtain the NRIO estimates.

The chlorophyll concentrations found on the Natal continental shelf fall within the range reported for the 'east coast of South Africa' by Krey (1973). They also agree closely with upper water-column values reported for North Carolina (Paffenhofer _et al_. 1980) and southern New York Bight (Falkowski _et al_.1983) continental shelf waters. Vertical structure in chlorophyll distribution is not as strongly developed off Natal as in the above localities, however, due to the lack of identifiable intrusions of bottom water onto the shelf.

The phytoplankton cell counts made by NRIO fall in the range of 0.1 to 25 x 10^4 cells/litre found by Turner (1984) in southeast USA continental shelf waters. This is lower than the values found over the ORI reef, and it is notable that Turner's samples were dominated by diatoms or, occasionally, dinoflagellates. Athecate flagellates were not abundant.

The biomasses recorded off Natal are, as expected, lower than those found in the upwelling region of the southern Benguela on the west coast of southern Africa. For the purposes of comparison chlorophyll-a concentrations in the latter region peak an order of magnitude higher than they do off Natal (e.g. Carter, 1982) and cell counts (diatoms) frequently exceed 10^6 cells/litre (De Decker, 1973).

Production

Phytoplankton production measurements in Natal coastal waters are not as extensive as the biomass measurements discussed above in that fewer estimates have been made and the areal coverage obtained is far more restricted. Further, there have been differences in the selection of sampling depths and thus not all of the results of the studies are comparable. Table 7.2 lists the measurements that have been made; integrated water column production estimates are given for only those studies that obtained adequate euphotic zone coverage.

Table 7.2 Means, standard deviations (SD), and ranges for phytoplankton production measurements made in Natal continental shelf waters (where only one measurement has been made it is listed under mean). Values published as hourly rates have been converted to values per 10-hour day for comparison purposes. The geographic locations of each of the studies referred to are depicted in Figure 7.1.

Source	Depth Range	Number of Measurements	(mgC/m^2/10 hour day) Mean	SD	Range
Ryther et al. (1966)	euphotic zone	4	1593	-	10-3140
Burchall (1968a)	"	2	592	-	243-942
Burchall (1968b)	"	92	513	386	32-2191
NRIO (unpub.)	"	1	805	-	-
Burchall (1968a)	oceanic waters off Durban	12	161	122	17-357
			(mgC/m^3/10 hour day)		
Oliff (1973)	10m	340	18	27	1-241
Schleyer (1981)	1m	29	128	85	17-313

Burchall's work (1968b) represents the most comprehensive coverage for the euphotic zone in that her primary production measurements were carried out over a six-year period, in which all months were covered. Geographic coverage, however, was limited in that only two fixed stations some 4 km apart on the Durban shelf were occupied. Despite the long measurement period, Burchall could not demonstrate any fixed relationships between volumetric or integrated water-column primary production rates and hydrographic features or nutrients in the form of phosphorus concentrations. Further, regression analysis of her tabulated data yields similar results for nitrate-nitrogen availability. This is probably because Burchall's data are inadequate in terms of time series coverage of specific forcing effects rather than the lack of any relationships.

A feature of Burchall's measurements are the deep euphotic zones encountered (mean depth = 43 m), and the uniform distribution of phytoplankton production in the water-column. This accords with the generally low phytoplankton biomasses recorded in Natal coastal waters. In this regard it is notable that for those stations with high integrated water-column production estimates (mean + 1SD), the mean euphotic zone depth was 48 m. The high estimates obtained, therefore, are more the result of moderate rates of production distributed over a deep-water column than of high individual volumetric values. The single NRIO measurement and the Burchall (1968a) estimates listed in Table 7.2 relate to similar situations.

Ryther et al. (1966), in the spring of 1964, carried out four phytoplankton production measurements spread over the wide central shelf. Three of these resulted in water-column estimates that are high (mean + 2SD) relative to Burchall's (1968b) results with one of them being the highest rate measured in Natal continental shelf waters to date. Ryther et al. attributed these high rates to terrigenous input of nutrients to coastal waters via the Tugela River. However, outflow from Natal rivers is small compared with those of their European and North American counterparts and appears to have little effect on the nutrient concentrations of even nearshore (5 km offshore) waters (e.g. Chapter 6). Further, Natal rivers carry high silt loads (Begg, 1978) so that strong outflows lead to high turbidities in the nearshore waters. Therefore, although nutrient concentrations may be increased under these conditions, primary production will be limited by low light penetration. It is thus more likely that the high rates measured were due to some other forcing event (e.g. upwelling), and in this context they must be viewed in conjunction with Burchall's (1968a) rate of 942 mgC/m^2/d measured south-east of the Tugela River in June, 1965. This will be considered further in the discussion at the end.

Oliff's (1973) mean volumetric production estimate is based on intensive measurements in Richards Bay shelf waters carried out over one-week periods in September and November, 1970, and April and October 1971. The mean is low and the variance high. The high variance can be attributed to the hydrographic variability of the region (Pearce 1977a)

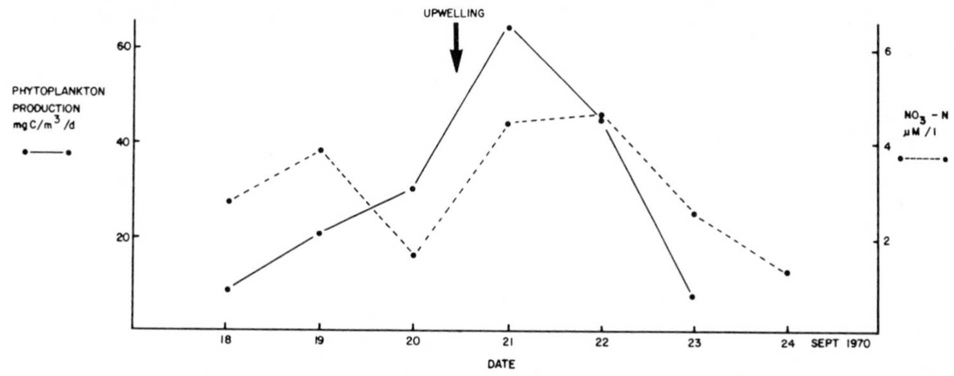

Figure 7.3 Changes in mean phytoplankton production rates and nitrate-nitrogen concentrations at 10 m depth in response to an upwelling even on the continental shelf off Richards Bay (adapted from Oliff, 1973).

and the varying light levels at 10 m depth due to varying turbidities caused by river discharge, etc. This latter factor would also have been instrumental in producing the low rates observed.

Oliff's primary production rates show the same lack of any fixed relationships with ambient nutrient concentrations as do Burchall's (1968b) data. However, the Richards Bay continental shelf is a region where coastal marine upwelling occurs (e.g. Chapter 6) and Figure 7.3 shows the response in production to one of these upwelling events. It can be seen that the peak in phytoplankton production coincides with the initial increase in nitrate-nitrogen concentration but lags the increase in phosphorus concentrations by one day. High production rates are very short-lived, declining within a day and reaching zero within two days of the upwelling event. This follows closely the pattern in phosphorus concentrations but nitrate-nitrogen concentrations were still greater than 2 µM/ℓ at this time. There are no apparent reasons for the decline in production. This lead and lag relationship with nutrients coupled to short term physical variability militates against the establishment of statistical relationships between the parameters. Oliff's (1973) data-listing provides other examples of similar short-term peaks in phytoplankton production off Richards Bay.

Schleyer (1981) obtained his measurements from a wave-washed nearshore reef (ORI reef) near Durban's central bathing beaches. The rates measured are high compared with Burchall's (1968b) surface

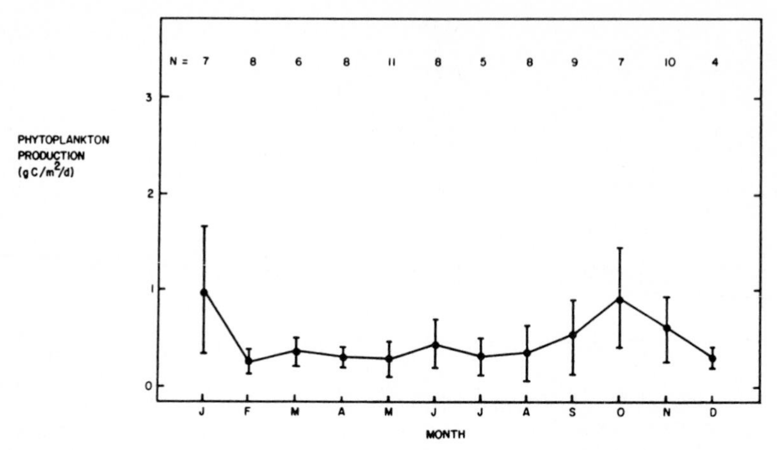

Figure 7.4 Mean monthly phytoplankton production rates measured off Durban for the period 1961-1966 (data from Burchall, 1968b).

measurements, and an order of magnitude higher than Oliff's (1973) 10m depth measurements off Richards Bay. These comparisons indicate that the nearshore waters off Durban are productive relative to shelf waters further offshore.

As was the case with Burchall (1968b) and Oliff (1973), Schleyer (1981) also found no consistent relationships with environmental variables. With intense microbial activity observed over the reef as well as the high biomass of benthic filter feeders on the reef itself (Shleyer, 1981; Berry and Schleyer, 1983), it is likely that nutrient regeneration rates are rapid. This would ensure adequate nitrogen concentrations, especially in the form of ammonia, to support the phytoplankton production rates observed.

The identification of seasonal patterns in phytoplankton production rates in Natal coastal waters is complicated by the few measurements made and the different locations covered. Burchall (1968b) provides the best data set and Figure 7.4 shows the distribution of the monthly means and standard deviations calculated from her tabulated data. There is an indication of a production maximum in the late (austral) winter/early spring period, but the high mean rate for January and the large variances cloud the picture.

A similar spring production peak occurred nearshore off Durban (Schleyer, 1981) and in Richards Bay continental shelf waters (Oliff, 1973), although in both cases there were large variances. A further pointer to a late winter/spring peak in phytoplankton production is the higher biomasses recorded off Durban in this period (Table 7.1). None of the above offers conclusive evidence of marked seasonal variations, but seasonality in the plankton will be pursued further below.

The primary production rates listed in Table 7.2 compare with the 380 and 522 mgC/m^2/day recorded in waters inshore of the Gulf Stream in the region of Cape Hatteras (Falkowski et al. 1983). The ranges are very large, however, with the peak rate exceeding the maximum of 3 000 mgC/m^2/day recorded for the New York Bight by Walsh et al. (1978). The annual water-column integrated phytoplankton production rate for Natal continental shelf waters calculated from Burchall's (1968b) data is 187 gC/m^2/year. This is larger than Ryther's (1969) estimate for the coastal zone, but lower than that for upwelling areas. The estimate is also considerbaly lower than the annual production rate of about 1130 gC/m^2/year obtained for the southern Benguela by Brown (1980) and Carter (1982).

Phytoplankton species distributions

Natal continental shelf waters have received little attention in terms of phytoplankton species distribution studies. The earliest records for the region are those of Shadbolt (1854) who, although concentrating on benthic flora, gave a description of a Bacteriastrum sp. Natal waters were then ignored until 1961 when the South African Division of Sea Fisheries published a list of 60 diatom species recorded during surveys carried out in 1959 and 1960. This was followed by Taylor's (1966) large-scale survey of the south-west Indian Ocean in which stations on the Natal continental shelf were included, and Thorrington-Smith's (1969) investigations of phytoplankton distributions on the continental shelf and in the Agulhas Current (see Figure 7.1 for locations). In the latter three studies Discovery type N50 nets were used (60 µm mesh aperture), hauled vertically through the top 100 m (S.A. Div. Sea Fish., 1961; Taylor, 1966) or 50 m (Thorrington-Smith, 1969). The studies were thus biased towards the larger forms in the phytoplankton although Thorrington-Smith used cell counts from water samples to supplement her net data.

Both Taylor and Thorrington-Smith found Natal continental shelf waters to be dominated by diatoms in terms of both number of taxa and numerical abundance. The latter author, for instance, recorded 115 diatom and 15 dinoflagellate taxa for this region. In general terms, all three of the above studies found similar species assemblages with similar dominant species. Table 7.3 presents a composite list of these species.

Table 7.3 Commonly occurring net phytoplankton species in Natal continental shelf waters as determined bt the South African Division of Sea Fisheries (1961), Taylor (1966) and Thorrington-Smith (1969).

Planktoniella sol	
Climacodium biconcavum	C. peruvianus
C. frauenfeldianum	C. radicans
Ditylum sol	Bacteriastrum elongatum
Eucampia cornuta	B. furcatum
Chaetoceros affinis	B. minus
C. atlanticus	Dactyliosolen mediterraneus
C. coarctatus	Guinardia flaccida
C. compressus	Rhizosolenia setigera
C. curvisetus	R. stolterfothii
C. eibenii	Thalassionema nitzschioides
C. lauderii	T. frauenfeldii
C. lorenzianus	T. longissima
C. messanensis	Nitzschia pacifica
C. pendulus	N. seriata

Thorrington-Smith (1969) computed similarity indices between three continental shelf stations and between these and oceanic stations on the basis of species composition. The results showed that most of the stations were relatively dissimilar despite the common dominating species. She also calculated species diversity indices (Simpson's diversity index, 1949) for all the stations sampled. Highest diversities were found for stations situated in the Agulhas Current; continental shelf and further offshore stations were characterized by lower diversities. Within this larger trend species diversities at the three continental shelf stations ranged from 0.06 to 0.11. The high (= low diversity) value was obtained from a station immediately off Durban which was dominated by <u>Chaetoceros lorenzianus</u>, <u>C. messanensis,</u> <u>Bacteriastratum varians</u> and <u>Thalassiothrix frauenfeldii</u>; these species comprised 70% of the total numbers. This appears to be an unusual situation in that the other stations, and those sampled by Taylor

(1966), had an admixture of oceanic species which led to higher diversities. The variation in phytoplankton species diversity between the shelf, Agulhas Current and further offshore stations is paralleled in copepod distributions (Carter, 1977).

Schleyer's (1981) cell counts indicate strong dominance by flagellates in ORI reef samples which differs from the diatom dominance evident in the above results. Schleyer explains this by the damaging effect of turbulence on the larger phytoplankton in the surf zone which would break up diatom chains. This explanation accounts for the fact that diatoms were few in number but suggests that there is little to moderate interchange between water in the surf zone and further offshore. This may not always be valid (e.g. Field et al. 1981). Parsons (1979) demonstrated that non-pulsed nutrient systems tend to support flagellates rather than diatoms, from which it can be inferred that the Durban bight area is remote from upwelling effects. Seawater temperatures measured off beaches in the area by various agencies (Sciocatti, unpublished) provide supporting evidence for this. This feature, combined with inferred high ammonia-nitrogen supply rates, contributes to the apparently unique species assemblage and high phytoplankton production rates reported by Schleyer (1981).

ZOOPLANKTON

As with the phytoplankton studies, zooplankton investigations in Natal coastal waters are not extensive and have concentrated upon biomass and species distributions. Systematic studies are few, most of the surveys having been either once-off or conducted with widely spaced stations. A further complicating factor is the lack of uniformity in sampling gear used, such as nets of differing dimensions, pumps, etc. Because of this it is difficult to make comparisons between the results of the surveys and thus these will be drawn in general terms only. The locations of the studies reviewed here are shown in Figure 7.1.

Biomass

Table 7.4 summarizes the published measurements of zooplankton biomass that have been made in Natal continental shelf waters. It is apparent that biomasses can attain moderate to high concentrations, especially in the inner shelf region, but are variable: this variability is strongly demonstrated in Carter's (1977) estimates for the Durban and Port Edward shelf regions. A consequence of this variability and the overall low numbers of samples is that the apparent regional differences, e.g. low biomasses off St Lucia, Richards Bay and Port Shepstone compared with those off Durban and Port Edward, must be treated with caution. Therefore, apart from the variability, the only conclusion that can be drawn from Table 7.4 is that zooplankton biomass in Natal continental shelf waters can be an order of magnitude greater than that further offshore in the Agulhas Current.

The large-scale variability in plankton biomass distributions evident in Table 7.4 has a smaller-scale counterpart, as was demonstrated by Carter (1973) in a systematic study of the smaller net plankton (Discovery N50 net, 60 µm mesh aperture) in continental shelf waters off Richards Bay. In this region biomass distribution patterns exhibited strong variation on time scales of days and spatial scales of tens of kilometres even when there were no corresponding physical variations in water column structure. Short-term variability notwithstanding, Carter was able to demonstrate a relationship between plankton biomass and current speed; low biomasses were associated with strong currents. Pearce (1973) showed that the strongest currents in the region were predominantly southward which implies that plankton biomasses north of the area are characteristically low. This is supported by the fact that the continental shelf north of Richards Bay is extremely narrow (e.g. Fig. 7.1), and is thus probably subject tofrequent incursions of Agulhas Current water which are low in plankton biomass (e.g. Table 7.4). The flushing effect demonstrated for the Richards Bay area is probably limited to narrow shelf regions e.g. off Port Shepstone and Port Edward, as these are the regions that can be directly affected by the Agulhas Current.

The only wider shelf region in which biomass was investigated in any detail was that off Durban where Carter (1977) studied zooplankton distributions on a transect across the Agulhas Current. Figure 7.5 depicts examples of the distributions found. As is evident in Table

Table 7.4 Summary of the published zooplankton biomass measurements made in Natal continental shelf waters. The data are listed as geographic series from north to south (see Figure 7.1); biomass estimates from 60-80 km offshore off Durban are included for comparison purposes. All estimates were derived from vertical hauls. (Inner shelf = depth 100 m, Outer shelf = depth 100-200 m). N is the number of samples.

Source	Location	Net	Haul Length (m)	Displaced vol ml/m^3 Mean	SD	Range	N
Zoutendyk & Sacks (1969)	St Lucia (Inner shelf)	N113 (IOSN)	60-0	0.08	-	0.07-0.09	2
Carter (1973)	Richards Bay (Inner shelf)	N113	50-0	0.05	0.02	0.02-0.08	13
Zoutendyk (1960)	Durban (Inner shelf)	N70 Discovery	40-0	0.47	-	0.18-0.96	3
Shipley & Zoutendyk (1964)	"	N70	40-0	0.63	0.45	0.24-1.10	4
Zoutendyk & Saks (1969)	"	N113	50-0	0.1	-	-	1
Carter (1977)	Outer shelf	WP II (SCOR)	180-0	0.23	0.16	0.07-0.76	18
Zoutendyk (1960)	Port Shepstone (Inner shelf)	N70	35-0	0.06	-	0.02-0,10	2
Carter (1977)	Port Edward (Inner shelf)	WP II	30-0	0.68	0.50	0.07-1.68	14
	(Outer shelf)	"	110-0	0.23	0.12	0.06-0.45	8
	Durban (Offshore)	"	200-0	0.06	0.02	0.04-0.10	18

7.4 the nearshore waters (10 and 20 km stations) support higher and more variable biomasses than do waters further offshore in the Agulhas Current, the boundary of which is clearly demarcated by strong gradients in surface temperature (Pearce, 1977b). Again there are short term variations in biomass levels (e.g. panels A and B), but in general the overall distribution patterns are stable, tending to break down only in the high surface temperature conditions which occur in late summer (Figure 7.5 H, I). These features are linked to changes in zooplankton population structure which will be discussed further below. The repeated single transect approach used in this study could not resolve longshore spatial variability, although other studies in the area suggest that this is marked (e.g. Burchall, 1968b; Thorrington-

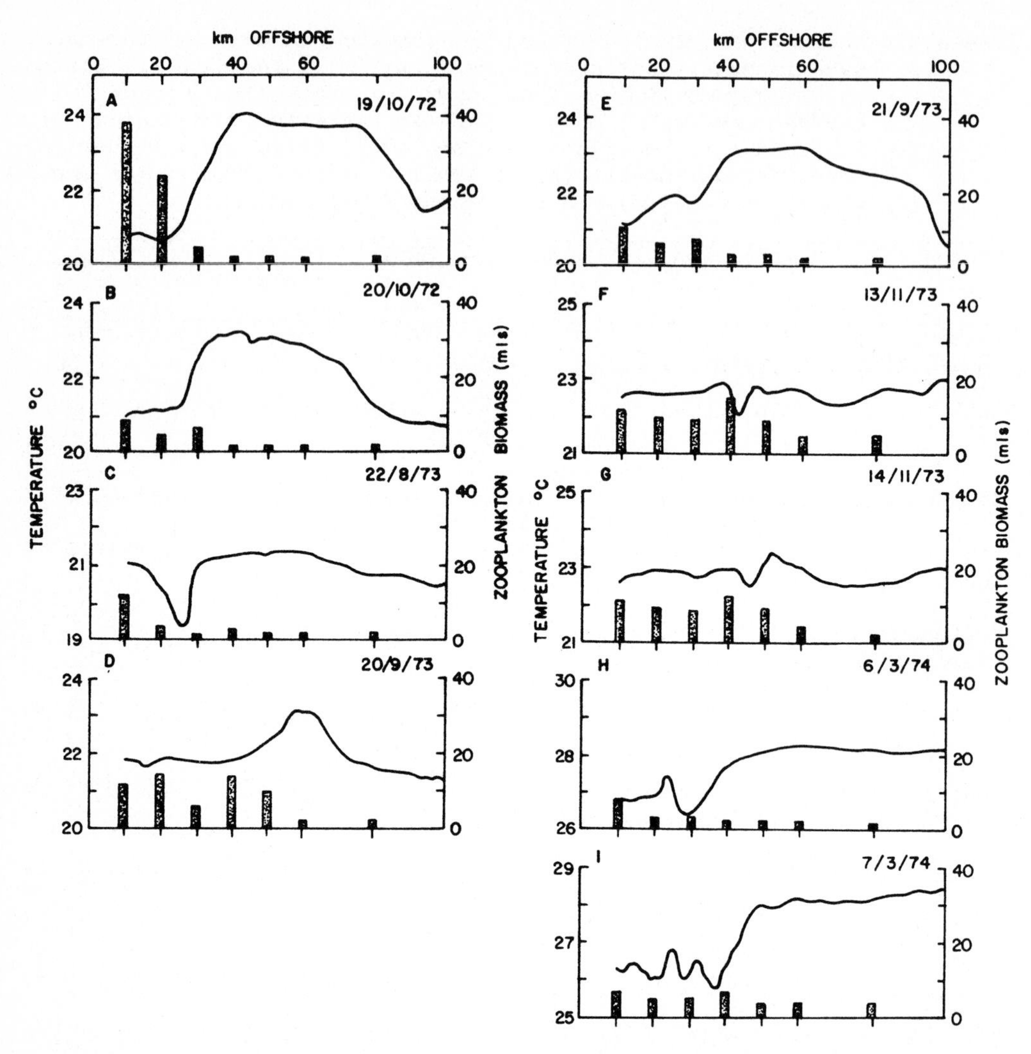

Figure 7.5 The distribution of zooplankton biomass (expressed as displaced volumes per standard WP2 200-0 m net haul (i.e. 50 m^3) across the Agulhas Current off Durban (after Carter, 1977).

Smith, 1969).

In a comparable study off Port Edward, which is characterized as a narrow continental shelf region (Chapter 5), Carter (1977) found biomass distribution patterns similar to those described for the Durban shelf (Figure 7.6). The major difference between the two regions is that the high biomasses characteristic of the nearshore persisted into the later summer, high-water temperature period. This feature will be

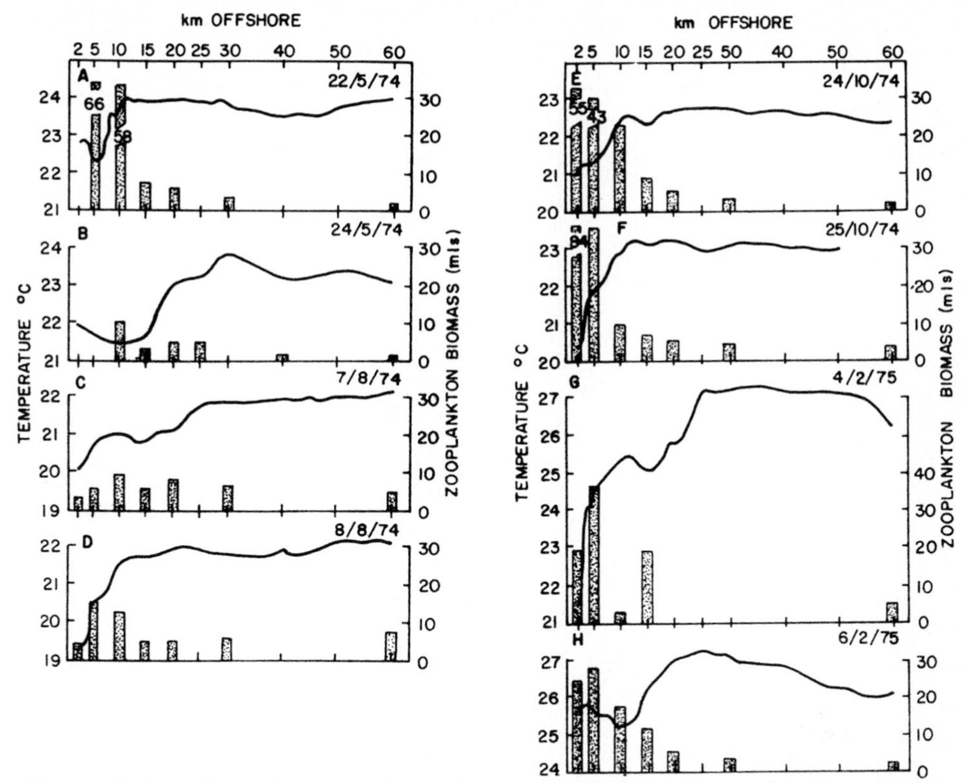

Figure 7.6 The distribution of zooplankton biomass (units as for Figure 7.5) across the Agulhas Current off Port Edward (after Carter, 1977).

discussed further in the section on zooplankton species distributions. Again no statement can be made about longshore spatial variability.

Information on the vertical distribution of zooplankton biomass in Natal continental shelf waters is extremely sparse, being limited to four depth-partitioned plankton net hauls (Shipley and Zoutendyk, 1964; Zoutendyk and Sacks, 1969) and seven (pumped) discrete depth samples from the region of the semipermanent cyclonic eddy off Durban (NRIO, unpublished).

The partitioned plankton net hauls indicate that the bulk (about 70%) of the zooplankton biomass is distributed in the top 100 m of the 180 to 220 m water column sampled; this finding was supported by 17 distributions determined for a 500 m-deep water column further offshore (NRIO, unpublished). The pump samples yield finer scale data over the top 50 m; the biomass profiles obtained are depicted in Figure 7.7. Five of the profiles show uniformly low concentrations over the

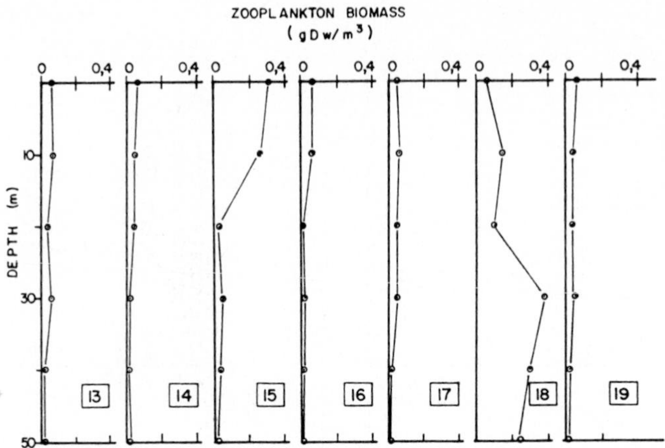

Figure 7.7 Zooplankton biomass profiles measured in the vicinity of the semipermanent cyclonic eddy off Durban during 31 August to 1 September, 1980; station numbers are given.

depth range sampled with the other two (15 and 18 August) showing completely opposite trends. In both cases the biomass peaks were due to greater numbers of the calanoid copepod species Paracalanus parvus and Centropages chierchiae. Temperature structure was invariant in this sampling period and although there were variations in chlorophyll-a distribution (Figure 7.2b), there is no apparent relationship between the two variables.

Gradients in the vertical distribution of zooplankton biomass in continental shelf waters are usually linked to strong physical gradients or the associated maxima in phytoplankton distributions (e.g. Paffenhofer, 1980; Hutchings, 1985). No such features have been identified in Natal shelf waters and thus the uniform distributions found in the limited surveys undertaken probably apply generally to the region.

In view of the large variability, the seasonal coverage obtained in the studies listed in Table 7.4 is poor. However, by combining data from Zoutendyk (1960) and Shipley and Zoutendyk (1964), biomass estimates for seven months of the year can be obtained for the inner shelf off Durban. No seasonal trends are evident from these data. Similarly, although Carter's (1977) Durban outer-shelf data (Figure 7.5) show a minimum in the high surface-temperature, late-summer period, this is not seen in the Port Edward data (Figure 7.6). This lack of strong, or even consistent, seasonal variation is also evident in plankton distributions off Richards Bay (Carter, 1973), where short-term physical variation was shown to be the controlling factor. This lack of seasonality in zooplankton biomass accords with the at-most weak seasonal variation in the phytoplankton levels, e.g. production (Figure 7.4), and absence of seasonality in shelf currents, upwelling, etc. (Pearce, 1977a).

The overall mean zooplankton biomass estimate for Natal continental shelf waters obtained from the data listed in Table 7.4 plust the NRIO unpublished data is 0.285 $m\ell/m^3$ (= 45.6 mg DW/m^3; Raymont, 1983). This is in the range 34 to 67 mg DW/m^3 found in the hydrographically similar North Carolina shelf waters (Paffenhofer, 1980), but is lower than the mean of 1.07 $m\ell/m^3$ quoted for the productive New York Bight waters by Raymont (1983) and the approximate estimate of 63 mg DW/m^3 for the surface layers of the seasonally upwelling-dominated, but physically variable, southern Benguela region (Andrews and Hutchings, 1980).

Zooplankton species distribution

The distribution of zooplankton species in Natal continental shelf waters has not been extensively studied, the investigations that have been carried out being limited to copepoda and chaetognatha. The earliers reports for the region are those of Cleve (1904) and Brady (1914, 1915) whose copepod species records are incorporated into a composite list in a South African Sea Fisheries (1961) report. Subsequently, De Decker (1964) presented species lists and distributional data based on surface pump samples and De Decker (1973, 1984) gives, among data of wider geographic coverage, some information on the copepod fauna of Natal coastal waters.

Carter (1977) and Schleyer (1976, 1985) carried out more systematic studies on calanoid copepod and chaetognath distributions on the transects across the Agulhas Current discussed above (Figures 7.5, 7.6). Carter identified a total of 186 copepod species from samples taken on the transects, all of which were recorded at stations inshore of the Agulhas Current boundary. Some of the species encountered were rare, e.g. Centraugaptilus horridus, Scolecithricella longipes, but 56 species occurred commonly enough in the shelf station samples to comprise one percent or more by number of the shelf species composition. These commoner species are listed in Table 7.5.

Carter (1977) was able to define neritic and oceanic species assemblages on the basis of Bray-Curtis similarity analyses (Bray and Curtis, 1957) of the distribution of copepod species across the transect. He found that the neritic assemblage had two components; one comprising those species that although reaching their highest numbers on the shelf were also observed in the Agulhas Current, and the other consisting of those species whose distributions were largely restricted to shelf waters. The dominant species of the former component was Paracalanus parvus, the other important species being Eucalanus mucronatus, E. pileatus and Acrocalanus gracilis. These two species dominated the copepod species composition, in terms of both adults and juveniles, of the high biomass samples found in the waters inshore of the Agulhas Current off Durban, as depicted in Figure 7.5. However, this was not the case in the high surface-temperature conditions during March, 1974 when biomasses were lower (Figure 7.5 H, I). In this period species characteristic of waters offshore of the western boundary of the Agulhas Current were found in the nearshore waters together with species of the unrestricted neritic component. Species diversity was consequently high (Carter, 1977; cf. Thorrington-Smith, 1969).

C. carinatus and C. chierchiae were also recorded in the high-biomass samples collected inshore of the Agulhas Current boundary off Port Edward. However, the two species dominated the assemblage only in the late spring period (Figure 7.6 E, F), the latter species being far more numerous in terms of both adults and juveniles. The absence of juveniles of the former species was marked. Species of the unrestricted neritic component, P. parvus in particular, dominated the shelf assemblage during the other periods.

The predominance of south-going currents on the Port Edward shelf (Chapter 5) suggests that the restricted neritic assemblage species recorded off Port Edward were transported downstream from the continental shelf area off Durban, which appears to be a local centre of distribution for these species. De Decker (1984) argues, from larger-scale distributional data, that the Agulhas Bank is the major

Table 7.5 Calanoid Copepoda which commonly occur in waters inshore of the western boundary of the Agulhas Current off Natal.

Calanus finmarchicus	Mecynocera clausi
C. tenuicornis	Clausocalanus arcuicornis
Nannocalanus minor	C. mastigophorus
Canthocalanus pauper	C. furcatus
Calanoides carinatus	C. paululus
Undinula vulgaris	C. pergens
U. darwinii	C. farrani
Eucalanus hyalinus	C. minor
E. crassus	C. parapergens
E. mucronatus	Ctenocalanus vanus
E. pileatus	Euaetideus acutus
E. subcrassus	Scaphocalanus curtus
E. subtenuis	Scolecithricella ctenopus
E. sewelli	Scolecithrix danae
Rhincalanus cornutus	S. bradyi
R. nasutus	Centropages furcatus
Acrocalanus gibber/longicornis	C. chierchiae
A. gracilis	Temora turbinata
A. andersoni	T. discaudata
Paracalanus parvus	Temoropia mayumbaensis
P. pygmaeus	Pleuromamma gracilis
P. aculeatus	P. piseki
Calocalanus pavo	Lucicutia flavicornis
C. plumulosus	Heterorhabdus papilliger
C. styliremis	H. spinifer
C. contractus	Haloptilus longicornis
	Acartia negligens
	A. danae

centre of dispersal for C. chierchiae along the south-eastern and eastern coasts of South Africa, and probably acts as a reservoir for the Natal population. This requires northward flow in shelf currents which, at most, appears to be sporadic on the lower south coast (Chapter 5). However, the presence of the notal species Calanoides macrocarinatus (Brodskii, 1975) in Natal continental shelf waters in winter and spring (Carter, 1977), together with seasonal migrations of pilchards, Sardinops ocellata, into Natal waters from the south (e.g. Baird, 1971) indicates that there may be more interchange between the two centres than is indicated by direct current measurements.

Schleyer (1976, 1985) reported on chaetognath distributions deduced from species counts carried out on the samples collected by Carter (1977) off Durban. The sampling strategy employed was not efficient for chaetognatha, as adults and larger forms probably avoided the comparatively small net (WP 2) that was used. This resulted in small numbers of adult chaetognatha in the samples, and to counteract the effects of this in the grouping index (Bray-Curtis) Schleyer included juveniles in his analyses. This resulted in some identification problems, so that and thus eight species had to be grouped into four pairs. Schleyer found that the chaetognath community off Natal was dominated by _Sagitta minima_, _S. enflata_, _S. serratodentata/pacifica_ and _S. regularis/neglecta_. Of these species _S. enflata_ and _S. serratodentata/pacifica_ dominated the nearshore region, although all species occurred throughout the transect. Schleyer (1976, 1985) found that the dominance of the nearshore region by _S. enflata_ and _S. serratodentata/pacifica_ broke down during March, 1974 when water temperatures were high (Figure 7.5 H, I). These two species can thus be compared to the unrestricted component of the neritic copepod community.

DISCUSSION

The studies reviewed above have shown that plankton dynamics and distributions in Natal coastal waters are more strongly physically than biologically mediated. In the narrower continental shelf regions, as immediately north of Richards Bay, coastal margin upwelling has a strong effect on phytoplankton production rates, increasing them greatly. This effect is transient, however, as the upwellings are generally of short duration. The biomass resulting from the increase in production appears to be either dissipated by the turbulence associated with strong currents or advected away from the locality of the upwelling.

In the wider Durban shelf region identified upwelling processes are limited to cyclonic eddies which, although frequent (e.g. Pearce _et al_. 1979), do not have dramatic effects. It is concluded from associated increases in nutrient concentrations in the euphotic zone (e.g. Chapter 6) that phytoplankton production rates are enhanced by these eddies. Zooplankton grazing coupled with the dissipative effects of turbulence militate against the development of large phytoplankton biomasses, and this leads to deep euphotic zones being maintained and thus to high areal phytoplankton production rates.

The few phytoplankton production measurements that have been made in the wide central shelf region indicate high rates. These are probably driven by coastal marine upwellings north of the Tugela River as shown by sea-surface temperature maps derived from airborne radiation thermometry (Snyman, 1969), and by nutrients advected into the area from eddy-centre upwellings off Durban. Central shelf waters are relatively remote from the Agulhas Current, and it is inferred that current speeds are low (see Chapter 5). This would allow the development of high phytoplankton biomasses and to a more efficient coupling with the zooplankton. This region has been neglected by research workers, however, and the above conclusions require substantiation.

It is inferred, on the basis of dominance of the phytoplankton by flagellates, that the nearshore region off Durban is remote from upwelling effects and that regenerated nitrogen (ammonia-nitrogen) is the nutrient source for phytoplankton production in the region.

The limited zooplankton studies off Natal show that the region supports moderate zooplankton biomasses. Seasonality is not evident in the biomass distributions but is reflected in the copepod species assemblages in dominance by _C. carinatus_ and _C. chierchiae_ in the winter/spring period. This coincides with periods of elevated primary production. _P. parvus_ largely dominates the shelf copepod assemblage in the other seasons, although there is a general increase in species numbers in summer. This latter feature is also evident in the chaetognath distributions.

To a large extent plankton production rates and biomasses in Natal coastal waters parallel those in surface waters inshore of the Gulf Stream off North Carolina. The latter area, however, is characterized by seasonal intrusions of cooler bottom water which markedly affect the plankton dynamics of the region. These have not been identified in Natal waters to date; this may be due to inadequate coverage in the central shelf region or to some dynamic dissimilarity between the two regions.

The variability of Natal continental shelf waters complicates the interpretation of the results obtained in the plankton investigations carried out to date. In such circumstances good time-series data are required to adequately resolve the important processes; this has not yet been achieved.

REFERENCES

ANDREWS, W R H and L HUTCHINGS (1980). Upwelling in the southern Benguela Current. **Progress in Oceanography,** 9(1), 1-88.

BAIRD, D (1971). Seasonal occurrence of the pilchard Sardinops ocellate on the east coast of South Africa. **Investigational Report Division of Sea Fisheries, South Africa,** 96, 1-19.

BEGG, G (1978). The estuaries of Natal. **Natal Town and Regional Planning,** Vol. 41, 657 pp.

BERRY P and M H SCHLEYER (1983). The brown mussel Perna perna on the Natal coast, South Africa: utilization of available food and energy budget. **Marine Ecology Progress Series,** 13, 201-210.

BRADY, G S (1914). On some pelagic Entomostraca collected by Mr J Y Gibson in Durban Bay. **Annals of the Durban Museum, Part I,** 1-9.

BRADY, G S (1915). Notes on the pelagic Entomostraca of Durban Bay. **Annals of the Durban Museum, Part. I,** 134-136.

BRAY, J R and J T CURTIS (1957). An ordination of the upland forest communities of southern Wisconsin. **Ecological Monographs,** 27, 325-349.

BRODSKII, J J (1975). Phylogeny of the family Calanidae (Copepods) on the basis of a comparative morphologi9cal analysis of its characters. In: **Geographical and seasonal variability of marine plankton,** (Ed.) Zh.A. Zvereva, Keter Press, Jerusalem, 1-127.

BROWN, P.C. (1980). **Phytoplankton production studies in the coastal waters off the Cape Peninsula, South Africa.** MSc Thesis, University of Cape Town, 98pp.

BURCHALL, J (1968a). Primary production studies in the Agulhas Current region off Natal - June, 1965. **Investigational Report, Oceanographic Research Institute, Durban,** 20, 16pp.

BURCHALL, J (1968b). An evaluation of primary productivity studies in the continental shelf region of the Agulhas Current near Durban (1961-1966). **Investigational Report, Oceanographic Research Institute, Durban,** 21, 44pp.

CARTER, R A (1973). Factors affecting the development and distribution of marine plankton in the vicinity of Richards Bay. South African National Oceanographic Symposium, Cape Town. CSIR/NPRL Oceanography Division Contribution No. 30, 34 pp.

CARTER, R A (1977). The distribution of calanoid copepods in the Agulhas Current system off Natal, South Africa. MSc Thesis, University of Natal, 165 pp.

CARTER, R A (1982). Phytoplankton biomass and production in a southern Benguela kelp bed system. **Marine Ecology Progress Series,** 8, 9-14.

CARTER R A, P D BARTLETT and V P SWART (1986). Estimates of the nitrogen flux required for the maintenance of subsurface chlorophyll maxima on the Agulhas Bank. In: **Ocean hydrodynamics,** Proceedings of the 17th International Liege Colloquium. (Ed: J C J NIHOUL).

CLEVE, P T (1904). The plankton of the South African seas. I. Copepoda. **Marine Investigations in South Africa, Part 3,** 177-210.

DE DECKER, A (1964). Observations on the ecology and distribution of copepods in the marine plankton of South Africa. **South African Division of Sea Fisheries Investigational Report,** 49, 33pp.

DE DECKER, A (1973). Agulhas Bank plankton. In: **The biology of the Indian Ocean,** (Ed.) B. Zeitschel. Springer-Verlag Berlin, 189-219.

DE DECKER, A (1984). Near surface copepod distribution the southwestern Indian Ocean and southeastern Atlantic Ocean. **Annals of the South African Museum,** 93(5), 303-370.

FALKOWSKI P G, J VIDAL, T S HOPKINS, G T ROWED, T E WHITLEDGE and W G HARRISON (1983). Summer nutrient dynamics in the Middle Atlantic Bight: primary production and utilization of phytoplankton carbon. **Journal of Plankton Research,** 5(4), 515-537.

FIELD J G, C L GRIFFITHS, E A S LINLEY, P ZOUTENDYK and R A CARTER (1981). Wind-induced water movements in a Benguela kelp bed. In: **Coastal upwelling,** Coastal and Estuarine Sciences 1. (Ed: F A RICHARDS). American Geophysical Union, 507-513.

GARDNER B D, A D CONNELL, G A EAGLE, A G S MOLDAN, W D OLIFF, M J ORREN and R J WATLING (983). South African marine pollution survey report 1976-1979. **South African National Scientific Programmes Report 73,** CSIR, Pretoria.

HOBBIE J E, R J DALEY and S JASPER (1977). Use of Nuclepore filters for counting bacteria by fluorescence microscopy. **Applied Environmental Microbiology,** 33, 1225-1228.

HOLLIGAN P M, W M BALCH and C M YENTSCH (1984). The significance of subsurface chlorophyll, nitrate and ammonia maxima in relation to nitrogen for phytoplankton growth in stratified waters of the Gulf of Maine. **Journal of Marine Research,** 47, 1052-1073.

HUTCHINGS, L (1985). Vertical distribution of mesozooplankton at an active upwelling site in the southern Benguela Current, December 1969. Sea Fisheries Research Institute, Department of Environment Affairs, Republic of South africa, **Investigational Report 129,** 67 pp.

KREY, J (1973). Primary production in the Indian Ocean 1. In: **The Biology of the Indian Ocean,** (Ed.) B Zeitschel, Springer-Verlag,

Berlin, 115-126.

LUNDIE, G S H (1979). Nimbus 7 - CZCS. East coast siltation project, 21-23 March, 1979. National Research Institute for Oceanology, Stellenbosch, R.S.A., Memorandum 7934, 21pp.

OLIFF, W D (1973). Chemistry and productivity at Richards Bay. **CSIR NPRL Oceanography Division Contract Report No CFIS 37B,** Durban, South Africa, 17 pp.

PAFFENHOFER, G-A (1980). Zooplankton distribution as related to summer hydrographic conditions in Onslow Bay, North Carolina. **Bulletin of Marine Science,** 30(4), 819-832.

PAFFENHOFER G-A, D DEIBEL, L P ATKINSON and E M DUNSTAN (1980). The relation of concentration and size distribution of suspended particulate matter to hydrography in Onslow Bay, North Carolina. **Deep-Sea Research,** 27A, 435-447.

PARSONS, T R (1979). Some ecological, experimental and evolutionary aspects of the upwelling ecosystem. **South African Journal of Science,** 75, 536-540.

PEARCE, A F (1973). Water properties and currents off Richards Bay. South African National Oceanographic Symposium, Cape Town. 28 pp.

PEARCE, A F (1977a). The shelf circulation off the east coast of South Africa. MSc Thesis, University of Natal, 266 pp.

PEARCE, A F (1977b). Some features of the upper 500 m of the Agulhas Current. **Journal of Marine Research,** 35(4), 731-653.

PEARCE A F, E H SCHUMANN and G S LUNDIE 81979). Features of the shelf circulation off the Natal coast. **South African Journal of Science,** 74, 328-331.

RAYMONT, J G (1983). **Plankton and productivity in the oceans. Zooplankton.** Pergamon Press. 824 pp.

RYTHER, J H (1969). Photosynthesis and fish production in the sea. The production of organic matter and its conversion to higher forms of life vary throughout the world ocean. **Science,** 166, 72-76.

RYTHER J H, J R HALL, A K PEASE, A BAKUN and M M JONES (1966). Primary organic production in relation to the chemistry and hydrography of the western Indian Ocean. **Limnology and Oceanography,** 11(3), 371-380.

SCHLEYER, M H (1976). Chaetognatha as indicators of watermasses in the Agulhas Current system. MSc Thesis, University of Natal. 27 pp.

SCHLEYER, M H (1979). Preliminary results of a comparative study of the roles of bacteria and phytoplankton in the littoral waters of Natal. **South African Journal of Science,** 75, 566.

SCHLEYER, M H (1981). Microorganisms and detritus in the water column of a subtidal reef of Natal. **Marine Ecology Progress Series,** 4, 307-320.

SCHLEYER, M H (1985). Chaetognaths as indicators of water masses in the Agulhas Current system. **Investigational Report, Oceanographic Research Institute, Durban,** 61, 20 pp.

SHADBOLT, G (1895). A short description of some new forms of Diatomaceae from Port Natal. **Transactions of the Microscopical Society, N.S.** 2, 13-18.

SHIPLEY A M and P ZOUTENDYK (1964). Hydrographic and plankton collected in the south west Indian Ocean during the SCOR International Indian Ocean Expedition 1962-1963. **University of Cape Town Oceanography Department Data Report No. 2,** 210pp.

SNYMAN, C G (1969). Radiation thermometry off the Natal coast during the year 1968. **CSIR/NPRL Oceanography Division Internal General Report,** IG 69/1, 10 pp.

SOUTH AFRICAN DIVISION OF SEA FISHERIES (1961). Fisheries research in Natal waters. CSIR Symposium S2. **Marine studies off the Natal coast.** Pretoria, South Africa, 89-117.

TAYLOR, F J R (1966). Phytoplankton of the southwestern Indian Ocean. **Nova Hedwigia** 12(3/4), 433-476.

THORRINGTON-SMITH, M (1969). Phytoplankton studies in the Agulhas Current region off the Natal coast. **Investigational Report, Oceanographic Research Institute, Durban,** 23, 24 pp.

TURNER, J T (1984). Zooplankton feeding ecology: contents of faecal pellets of the copepods Acartia tonsa and Labidocera aestiva from continental shelf waters near the mouth of the Mississipi river. **Marine Ecology,** 5(3), 265-282.

WALSH J J, WHITLEDGE T E, BARVENICK F W, C D WIRICK and S O HOWE (1978). Wind events and food chain dynamics within the New York Bight. **Limnology and Oceanography,** 23, 659-683.

ZAR, J H (1974). **Biostatistical analysis.** Prentice-Hall, Englewood Cliffs, N J, U.S.A. 620 pp.

ZOUTENDYK, P (1960). Hydrographic and plankton data collected in the Agulhas Current during I.G.Y. **University of Cape Town Oceanography Department** 1, 89 pp.

ZOUTENDYK, P (1970). Zooplankton density in the southwestern Indian Ocean. Symposium on Oceanography in South Africa, Durban, 1970. 13 pp.

ZOUTENDYK P Aand D SACKS (1969). Hydrographic and plankton data, 1960-1965. **University of Cape Town Oceanography Department Data Report No.** 3, 82 pp.

Chapter 8

BENTHOS OF THE NATAL CONTINENTAL SHELF

T P McClurg
National Institute for Water Research
Council for Scientific and Industrial Research

DEFINITION AND SCOPE

The marine benthos comprises all organisms, plant and animal, which live on, or are closely associated with, the sea bed. Within this broad definition several sub-groups are recognized. These include phytobenthos (plant material) and zoobenthos (animal material). Larger organisms, which would be retained by a 1 mm mesh sieve, are referred to as macrobenthos, while those which pass through a 1 mm sieve but are retained by a 0.045 mm mesh sieve are referred to as the meiobenthos. Finally, microbenthos comprises that component which passes through a 0.045 mm mesh sieve, and consists essentially of diatoms, bacteria and protozoa. The meiobenthos can be further sub-divided into the temporary meiobenthos, which includes the juvenile stages of the macrobenthos, and the permanent meiobenthos, which comprises mainly rotifers, gastrotrichs, kinorhynchs, nematodes, archiannelids, tardigrades, harpacticoid copepods, ostracods, mystacocarids, turbellarians and oligochaetes.

These groups of marine benthos have been erected mainly on the grounds of convenience as each category requires specialised study techniques. However, in broad ecological terms they must be regarded as a continuum. They share the same habitat, they interact and are subject to the same environmental pressures.

In this chapter an attempt is made to present an overview of present knowledge of the marine benthos inhabiting the continental shelf off Natal. Attention will be focused on the region beyond the surf zone. The surf zone itself has been neglected, but is now receiving some attention from scientists at the Oceanographic Research Institute in Durban (Anon, 1984).

EARLY TAXONOMIC COLLECTIONS

Systematic biological work on the continental shelf off Natal started at the beginning of this century when the steam trawler Pieter Faure was loaned to the Natal Government to survey potential fishing grounds. Until then only a limited amount of collecting had been carried out by casual visitors and professional naturalist collectors.

The 19th century had seen a burgeoning of both amateur and professional interest in natural history. Professional collectors were commissioned by institutions or wealthy patrons to make collections of natural history objects around the world; the European preoccupation with conchology in the latter half of the century was symptomatic of this trend. Although molluscs were a prime target, all marine species were of interest. In southern Africa most of the professional collectors confined their efforts to the Cape but several, notably C F Krauss, J A Wahlberg and A Delegorgue, travelled north-eastwards as far as Natal. These early collections were restricted to material from the shore and so, in fact, contributed relatively little to knowledge of benthos beyond the surf zone. Nevertheless, a number of offshore species were inevitably included in these collections to be subsequently described by European taxonomists.

During the latter half of the 19th century and the early part of the 20th century, a number of great marine exploring expeditions were undertaken. These included the well-known Challenger and Discovery expeditions. Many of these expeditions paid passing visits to the Cape of Good Hope and some spent a few days at Port Natal. However, no systematic collections of note were made of benthos off the Natal coast.

SYSTEMATIC SURVEYS

At the turn of the century the Cape Government appointed Dr J D Gilchrist to take charge of the Fisheries and Marine Biological Surveys. The steamship Pieter Faure was acquired for this purpose, with initial emphasis on Cape waters. In 1900 to 1901 a special survey was made of the Natal fishing grounds on behalf of the Natal

Government, and Barnard (1964) provides a description of this work. In 1921, the Pieter Faure was replaced by the Pickle and the surveys were continued. Much of the material collected by the Pieter Faure was sent to overseas experts and the results were published in the volumes of Marine Investigations in South Africa. Gilchrist was appointed to the chair of Zoology at the South African College (later to become the University of Cape Town) in 1907, and an enduring interest in marine biology began at the University. This was nurtured and expanded by succeeding Professors of Zoology to encompass the whole coast of southern Africa.

Gilchrist was succeeded by Dr Keppel Barnard at the South African Museum in Cape Town. Barnard was a prolific taxonomist and published over 200 papers and three monographs (fishes, decapod crustaceans and molluscs). On this strong foundation a good basic knowledge of marine fauna in southern Africa was established. Nevertheless, knowledge of marine life on the east coast, particularly in the deeper water, tended to lag behind. This was because the area was geographically isolated and did not appear to offer much in the way of commercial fisheries. An attempt to rectify this imbalance was initiated by the South African Museum in the 1970s. With financial assistance from the South African National Committee for Oceanographic Research they completed five biological surveys off the east coast from 1975 to 1979 using the CSIR research vessel Meiring Naudé (Louw, 1977, 1980). Material from these cruises was referred to relevant specialists for description. Results that have so far been published in the Annals of the South African Museum include descriptions of decapod crustaceans (Kensley, 1977a, 1977b, 1981a), hydroids (Millard, 1977, 1980), echinoderms (Clark, 1977), amphipods (Griffiths, 1977), isopods (Kensley, 1978c, 1978a, 1984), bryozoans (Hayward and Cook, 1979, 1983) and myctophid fish (Hulley, 1984). Descriptions of the brachiopods will be published soon (Hiller, in press). In addition, the South African Museum has recently launched a study of the hitherto neglected soft coral fauna (Octocorralia) of southern Africa (Louw, pers. comm.).

Since 1980 the Natal Museum has undertaken annual cruises using the R V Meiring Naudé with the objective of extending the knowledge of marine mollusca on the east coast. Molluscs collected on these cruises are being examined and described by Dr R N Kilburn while some of the non-mollusc material is being referred to the South African Museum. Despite the recent attention that has been devoted to the benthos, knowledge of certain groups remains scant. Table 8.1 summarizes the state of knowledge of the major groups of benthic invertebrates. No attempt has been made to list all relevant literature but some key references have been cited to provide entry points. This assessment is inevitably subjective as it relies on speculative estimation of which groups will yield more species as collecting is extended to new habitats. A little over half of the groups are considered to be well known taxonomically, with the remainder being classified as poorly to moderately well known. To this latter group can be added the whole suite of invertebrates which comprise meiofauna. A start has been made on the intertidal meiofauna but this work has yet to be extended to the deep sea. Algae from the continental shelf beyond the surf zone are also poorly known. A limited knowledge has been gained from deep-sea specimens which have been cast ashore but a systematic study has yet to be made.

The available knowledge of many of the benthic groups is thus too sparse to allow definitive discussion on the zoogeography of the benthos. However, in describing components of the Natal benthos, most specialist taxonomists have made tentative zoogeographical assessments. Before these assessments are reviewed, it is necessary to consider the physical factors which might influence the survival and distribution of benthos.

PHYSICAL FACTORS

Although the distribution of Natal's intertidal fauna in relation to physical and hydrographic factors has been well documented (see Stephenson and Stephenson, 1972), relatively little has been published on the distribution of shelf fauna. A notable exception is the work of Heydorn and co-workers (1978) who have reviewed the ecology of the Agulhas current region with an emphasis on biological responses to environmental factors. This work highlights the poor state of knowledge of the east coast shelf benthos and identifies the Agulhas current, and the presence of numerous rivers entering the sea, as

Table 8.1 An assessment of the present state of knowledge of major invertebrate taxa on the Natal continental shelf.

Group	State of Knowledge	Key References
Actinaria	Poor	Carlgren 1938
Amphipoda	Good	Griffiths 1974, 1976, 1977
Anomura	Moderate	Barnard 1950; Kensley 1977a, 1981a
Brachiopoda	Good	Hiller (in press)
Brachyura	Good	Barnard 1950; Kensley 1977a, 1981a
Bryozoa	Good	Hayward and Cook 1979, 1983
Cirripedia	Good	Barnard 1924a
Cumacea	Good	Day 1975, 1978a, b, 1980
Echinodermata	Good	Clark 1977, Clark and Courtman-Stock 1976; Thandar 1984
Echiurida	Moderate	Wesenberg-Lund 1963; Stephen and Cutler 1969; Webb 1972; Biseswar 1985
Hydroida	Good	Millard 1958, 1959, 1975, 1977, 1978, 1980
Isopoda	Good	Kensley 1978b, c, 1984
Macrura	Good	Barnard 1950; Kensley 1972, 1981a; de Freitas 1980, 1984a, b; 1985a, b
Madreporaria	Poor	Crossland 1948; Boschoff 1981
Mollusca	Good	Kilburn and Rippey 1982
Nemertea	Poor	Wheeler 1940
Octocorallia	Poor	Tixier-Durivault 1954
Porifera	Poor	Borojevic 1967; Levi 1963, 1967
Pycnogonida	Poor	Barnard 1954
Polychaeta	Good	Day 1967
Sipunculida	Moderate	Stephen 1942; Wesenberg-Lund 1963
Stomatopoda	Moderate	Barnard 1950
Tanaidacea	Moderate	Barnard 1940; Brown 1956, 1957

distinctive features of the south-east coast of Africa. These two factors have a far-reaching influence on the survival and distribution of benthic organisms.

The Agulhas current has already been described (Chapter 5). Its most obvious influence on the Natal shelf fauna is that of transporting tropical and subtropical species southwards. It provides a supply of larvae for potential settlement and ameliorates conditions for their survival. The Agulhas current is also the most important factor controlling sediment transport on the south-east African continental shelf (Chapter 3). In turn, sediment type is probably the most important factor in determining the survival and distribution of benthos. The Agulhas current clearly plays a dominant role in shaping the benthic ecosystem off Natal.

The draining of many rivers into the sea off Natal also has an important influence on the shelf benthos. Steep gradients exist in the relatively small catchment between the Drakensberg mountains (up to 3 000 m above sea level) and the sea (Figure 8.1). This, combined with a high annual rainfall (over 1 000 mm in places), means that soil erosion is an ever-present problem, exacerbated during the past century by high population growth and an increase in agricultural activities. Large quantities of sediment are discharged into the sea off Natal each year. Much of this is in the form of silt and clay which often causes extensive discoloration of the coastal waters (see Figure 5.9). Apart from reducing primary production because they decrease light penetration, high loads of suspended solids also adversely affect certain categories of benthos. Coral reefs and their associated flora and fauna are perhaps the most susceptible. There are no true coral reefs off the Natal coast, the most southerly being the fringing coral reef at Inhaca island off Maputo in southern Mozambique (Boshoff, 1980). However, in the area between St Lucia and the Mozambique border no major rivers enter the sea, and most of the offshore reefs are topped with prolific growths of coral. South of the St Lucia estuary, the abundance of corals diminishes rapidly, although good growths are evident on the Aliwal shoals south of Durban and isolated specimens have been recorded as far south as Port Elizabeth. It is difficult to evaluate the role of suspended solids in reducing coral growth on the Natal coast since other factors, such as lower water temperature and greater wave energy, must also play a role.

STUDIES ON THE MAIN BENTHIC INVERTEBRATE GROUPS

This section will review the findings of specialist taxonomists regarding the composition of the fauna and its zoogeographical affinity. To achieve this it will often be necessary to leave the geographical confines of the Natal coast and to consider the whole southern African coastline in relation to the tropical Indo-West Pacific region.

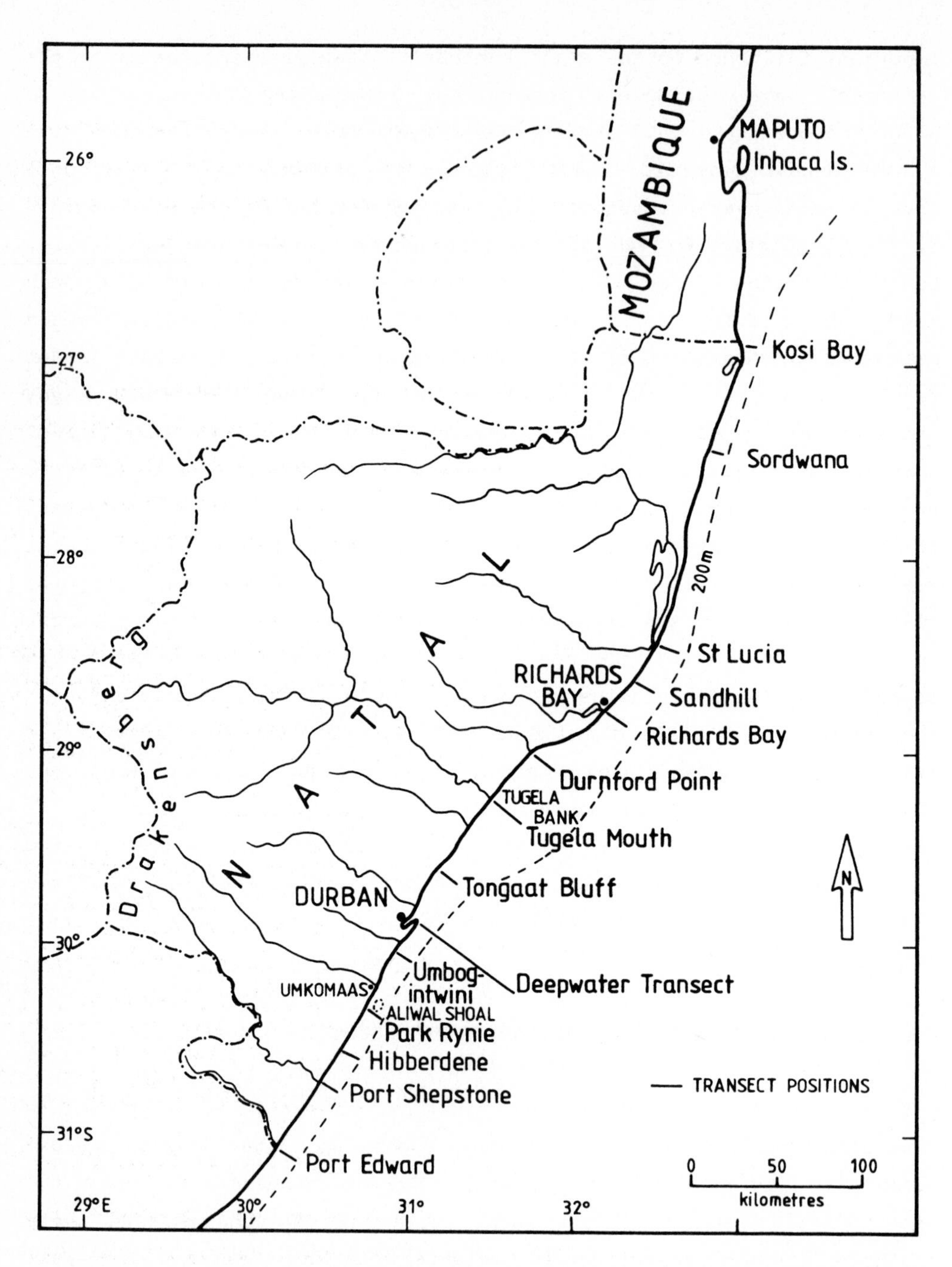

Figure 8.1 Map of the Natal coast showing physical features and the transect positions for marine pollution monitoring.

Amphipoda

Barnard (1940) listed about 200 species of gammarid and caprellid in his review of amphipod records for southern Africa. At the time of the next major review (Griffiths, 1976) this number had risen to about 300, of which 122, representing some 25 families, had been found off Natal. Approximately 35% of the species found off Natal were considered by Griffiths (1976) to be endemic to southern Africa, while a further 33.5% were found to have a tropical or sub-tropical affinity (12% Indo-Pacific, 18% Circumtropical, 2.5% S.W. Indian Ocean, 1% Indian). The remainder consisted mostly of Atlantic (14%), Cosmopolitan (8%) or Southern (7%) species. The Natal continental shelf seems therefore, to support a diverse assemblage of amphipods with a high degree of endemism and a marked tropical influence.

The amphipods collected in deeper waters off Natal by the R.V. Meiring Naudé in 1975 and 1976 were described by Griffiths (1977). This collection was small yet it yielded two undescribed genera and three new species records for southern Africa. It would seem that many more amphipod species await discovery in the deeper waters off Natal.

Brachiopoda

Although the fossil record shows that brachiopods were abundant in pre-historic seas, they are relatively rare today. Very little attention has been devoted to modern southern African brachiopods. Jackson (1952) in reviewing past research, listed a total of 15 species for the entire coastline between Saldanha Bay and Maputo, and Cooper (1973a, b, 1981) added a few more species to the list. The most significant work, as far as the east coast and Natal are concerned, is the description by Hiller (1986) of the brachiopods taken by the R.V. Meiring Naudé from 1975 to 1979. This material yielded 17 species, of which three were new to science and five represented new finds in southern African waters. Most of the specimens were taken in deep water (>500 m). This fauna showed a predictably strong affinity with other Indian Ocean faunas. However, two of the new species belonged to genera which had been recorded only in the Philippines and Caribbean. It appears, therefore that there are also many more unidentified brachiopod species off Natal, particularly in deeper waters.

Bryozoa

Early research on bryozoans in southern Africa (O'Donoghue, 1957, O'Donoghue and de Watteville, 1937, 1944) dealt almost entirely with specimens taken intertidally or from ships' hulls. The collections by the R.V. *Meiring Naudé* off the east coast from 1975 to 1979 and described by Hayward and Cook (1979, 1983) constitute the first comprehensive study of this group in the deeper waters of southern Africa. This is reflected in the very high number of new species recorded.

In describing the specimens from shallower waters (mostly < 100 m) Hayward and Cook (1983) recognized 130 species, of which one genus and 44 species were considered to be new. In the deeper shelf waters (> 350 m) Hayward and Cook (1979) discovered one new family, three new genera and 23 new species out of a total of 51 identifiable species. The discovery of so many new species was, according to Hayward and Cook (1979), not unexpected and echoed a similar recent experience in the north-eastern Atlantic. It does, however, emphasize how much work remains to be done, world-wide, on the Bryozoa.

Hayward and Cook (1983) remark on the strong similarities that exist between South African Bryozoa and those from Australian and northern New Zealand waters and suggest that this link may be of considerable antiquity. They regard the eastern South African bryozoan fauna as having a distinctive endemic element and a marked affinity with the Indo-West Pacific region. Much work is still required to fully characterize the bryozoan fauna of Natal, but this recent work must place the Natal Bryozoa amongst the better known in the world.

Cirripedia

The only detailed description of South African barnacles is that by Barnard (1924a) who listed 74 species for the entire country, comprising 40 stalked, 32 sessile and two parasitic forms. Many of these were recorded off Natal, having been collected by the *Pieter Faure* in 1901 and 1902. Barnard's list appears to be comprehensive and it is unlikely that many more species await discovery. The propensity for cirripedes to attach to moving objects such as turtles, logs and ships, weakens any zoogeographical discussion, so this will not be attempted.

Cumacea

Prior to the detailed study by J A Day (1975, 1978a, b, 1980) the cumacean fauna of southern Africa was poorly known, the only notable work as far as the east coast is concerned being that of Stebbing (1910, 1912). Day examined some 8 500 specimens from the whole of southern Africa and recognized about 80 species representing the families Bodotriidae, Lampropidae, Ceratocumatidae, Gynodiastylidae and Diastylidae. Of these, the Bodotriidae were the most diverse and abundant and dominated in the shallow water samples. The Diastylidae were also abundant but contained far fewer species and tended to be more abundant in deeper waters (> 200 m). The Lampropidae contained even fewer species which were restricted to deeper waters (> 200 m). The Gynodiastylidae showed a diversity and abundance similar to that of the Lampropidae, but were largely restricted to the shallower, warmer waters. The Ceratocumatidae was represented by a single species collected in deep water (650 to 900 m) off Natal.

Of the roughly 80 cumacean species recognized by Day for the whole of southern Africa, some 35 were recorded from Natal. This suggests that a relatively rich and diverse endemic cumacean fauna exists there. To what extent this impression may be due to uneven sampling effort can be ascertained only by additional surveys. Indeed, considerable further research is required, particularly in tropical waters north of Natal, before meaningful zoogeographical conclusions can be drawn.

Decapoda

Decapod crustaceans, which include the Anomura (hermit crabs, burrowing prawns), Brachyura (crabs) and Macrura (shrimps, prawns, lobsters) are, by virtue of their prominence and possible commercial value probably the best investigated of all the benthic invertebrates off the Natal coast.

After Barnard's (1950) monograph on the decapods no single comprehensive work appeared until the review by Kensley (1981b). Much taxonomic work had been done in the interim, for the number of recognized species, including terrestrial and freshwater forms, had risen from approximately 500 in 1950 to about 700 in 1981. Much of this increase arose from material which was collected by the R.V. Meiring Naudé off the east coast from 1975 to 1979 and subsequently described by Kensley (1977a,b, 1981a).

Certain species with exploitation potential have been the subjects of more detailed study. Berry (1969) investigated the distribution and biology of *Nephrops andamanicus* (langoustine or king prawn) on trawling grounds in deep water (around 500 m) off Natal. His emphasis was on population characteristics, morphometry, moulting and reproduction. Berry (1973) also made a similar study of *Palinurus delagoae*, a deep water spiny lobster of commercial significance, off the east coast. He took samples at depths between 180 and 324 m in an area extending from the Tugela River to Isipingo, just south of Durban. Results from this study were used in a subsequent taxonomic revision of the genus *Palinurus* in the south-western Indian Ocean by Berry and Plante (1973).

Haliporoides triarthrus, the pink prawn or knife prawn, also occurs in commercially significant quantities in deep water (180 to 650 m) off Natal but has, as yet, been the subject of only a brief exploratory study (Berry *et al*. 1975). Prawns inhabiting the shallower waters (<50 m) have, in contrast received considerable attention. Of particular interest is the work of de Freitas (1980, 1984a,b, 1985a,b,c) who made a detailed study of the superfamily Penaeoidea of south-east Africa. While the main thrust of this work was taxonomic, some attention was devoted to the general biology and ecology of the more economically important species. The penaeidae are by far the most prominent of the penaeoids in Natal and include commercially exploitable species such as *Penaeus indicus, P. monodon* and *Metapenaeus monoceros.* These species are characterized by the fact that their post-larval and juvenile forms inhabit estuarine and coastal backwater nursery areas, whereas the sexually mature adults live in the open sea to a depth of about 50 m. Penaeid prawns often comprise a prominent faunal component in Natal's estuaries and at St Lucia and Richards Bay they are sufficiently abundant to be commercially exploited as bait. The prawns of Natal's estuaries, and particularly Lake St Lucia, have been a subject of research for about 20 years (Joubert and Davies, 1966, Champion, 1976, Forbes and Benfield, 1985) while McClurg (1984) has made a preliminary study of the effects of some pollutants on P. indicus in relation to industrial development at Richards Bay.

The Agulhas current undoubtedly plays an important role in the dispersal of prawns on the Natal coast. MacNae (1962) recognized that the prawns of Natal were "derived from larvae washed in from further north". Champion (1970) suggested that in addition to this southward recruitment from Mozambique, a localized migration/recruitment cycle might operate in the Tugela Banks region. Kensley (1981b), while conceding that much taxonomic work remains to be done, reviewed the southern African decapod crustacean fauna on a zoogeographical basis. He concentrated his discussion on the species occurring in relatively shallow water (<200 m), a group which currently comprises over 80% of known species. For the southern African fauna as a whole 65.7% of the species had Indo-Pacific affinities, 19.6% were considered to be endemic, 5.7% had affinities with the Atlantic/Mediterranean and the remainder (9%) comprised all other categories (austral, widespread, uncertain). The Indo-Pacific component thus constitutes the major section of the shallow water (<200 m) decapod crustacean fauna. This component showed an expected increase from west to east along the coastline, with the percentage compositions being 12.9, 19.7, 77.0 and 93.0 for Port Elizabeth, East London, Durban and Maputo, respectively. The warm southward-flowing Agulhas current is the obvious mechanism for this marked Indo-Pacific influence. Kensley (1981b) also found a strong endemic element in the Natal decapod crustacean fauna, centred around Durban; 46.2% of southern African endemics were present in comparison to 38.5% and 24.0% for East London and Maputo, respectively.

Echinodermata

The Natal continental shelf supports a wide diversity of echinoderms. About 44% of the 280 species of crinoid, asteroid, ophiuroid and echiuroid described by Clark and Courtman-Stock (1976) as occurring in southern Africa (south of 23.5 °S) have been found off Natal. The majority of the species recognized by Clark and Courtman-Stock (1976) are either endemic to southern Africa (48%) or have affinities with the tropical Indo-West Pacific/Western Indian Ocean (30%), the Natal fauna being particularly well represented by the latter. The material collected by the R.V. Meiring Naudé in deep water (500 to 1 300 m) off Natal and described by A M Clark (1977) yielded 18 new records for South Africa (including one holothurian). It would thus appear that many more species await discovery in the deeper waters.

The holothurians were generally excluded from the work of A M Clark and had until recently received relatively little attention. H L Clark (1923) included them in his general review of South African echinoderms, whereas Deichmann's (1948) review was devoted entirely to the holothurians. Cherbonnier (1952, 1970), described further species from southern Africa. More recently, Thandar (1971, 1977, 1984, 1985) has studied holothurian collections made by the South African Museum and the Universities of Cape Town and Durban-Westville. On the basis of these collections, which comprised nearly 3 000 specimens, and the earlier literature, he provided a detailed description of holothuria from the whole of southern Africa. He recognized 122 nominal species. In considering their zoogeography Thandar (1984) discounted rare species and concentrated on the distribution of 68 species for which relatively comprehensive records are available. He found that the endemic and Indo-Pacific components comprised the bulk of known southern African holothuria, having contributed 41.2% and 45.6% of the species, respectively. The remainder comprised Cosmopolitan (7.4%), Circumtropical (2.9%) and Atlantic (2.9%) species. The Indo-Pacific component showed a predictable increase north-eastwards, comprising only 5.3% and 4.75% of the West Coast (Temperate) and South Coast (Temperate) fauna but 56.3% and 77.8% of the East Coast (Subtropical) and East Coast (Tropical) fauna, respectively.

The Natal coast fauna, which is included in the East Coast (Subtropical) component, is thus heavily influenced by the Indo-Pacific region. In terms of bathymetric distribution Thandar (1984) found that nearly 80% of known southern African species were restricted to shallow water (intertidal to 100 m). He observed that no holothurians had yet been found on the east coast in the 36 to 400 m depth range and suggested a number of reasons for their apparent absence. These included the relative narrowness and steepness of the continental shelf, the uneven topography of the sea bed, which in places hinders sampling, and massive sediment transport which may result in a physically unstable substratum. Additional surveys are clearly required to resolve the matter.

Hydroida

The hydroid fauna of southern Africa has been studied extensively, chiefly through the monographic work of Millard (1975). Papers that have dealt specifically with Natal hydroids include those by Millard in 1958, 1959, 1977 and 1980. The former two describe material arising from diverse early collections, and the latter two are devoted to collections made off the east coast by the R.V. Meiring Naudé. Millard (1978) has analysed the hydroid fauna of southern Africa (south of 20°S) on a zoogeographical basis. Of the 251 species considered, 182 or 72% have been found on the east coast (taken by Millard to include the Natal coast plus southern Moazambique). Of these, 72 (39.6%) were considered to be tropical, compared with 47 (25.8%) endemic, 12 (6.5%) temperate and 51 (28%) unclassified. There has clearly been a heavy invasion of hydroid species from the Indo-West Pacific region to the Natal coast through the influence of the Agulhas current.

Isopoda

The review of southern African marine isopods by Kensley (1978b) listed about 275 species, including parasitic forms for the whole continental shelf from abyssal depths to the shore. Of these, 48 non-parasitic species were recorded from beyond the intertidal zone on the Natal coast. This review included some of the non-anthurid isopods collected by the R.V. Meiring Naudé in deep water off the east coast and described by Kensley (1987c). Subsequent papers by Kensley (1978d, 1982, 1984) have dealt with the remaining isopod material taken by the R.V. Meiring Naudé.

Kensley (1984), in discussing the distribution and zoogeography of the 81 species taken by the R.V. Meiring Naudé between the southern Mozambique border and East London, concluded that the east coast supports a large endemic isopod fauna.

The Bathynatalidae are notable amongst the Natal isopods in that they represent the only endemic isopod family in southern Africa. The family was first recognized by Kensley (1978d) and now contains two genera and species, notably _Bathynatalia gilchristi_ which was first described by Barnard (1957) and _Naudea larvae_ which was described by Kensley (1979). As the family name implies these species occur in deep water (850 to 888 m off Natal).

Another isopod of note from Natal is the giant cirolanid _Parabathynomus natalensis_ which attains a length of over 8 cm. This species was first described by Barnard (1924b) from a specimen taken in 1920 from the R.V. _Pickle_ in 766 m of water off Natal. A second specimen was collected in 1964 by the _Anton Bruun_ in about 250 m of water off southern Mozambique, and Kensley (1978d) subsequently redescribed the species. The Natal continental shelf apparently supports a diverse and interesting isopod fauna.

Mollusca

Kilburn and Rippey (1982) have provided a zoogeographical analysis of southern African marine molluscs. They restricted their analysis to the littoral region and included only intertidal species and those in sufficiently shallow water to wash up on the shore. Lesser-known micro-molluscs and nudibranchs were intentionally excluded and the final number of species considered was about 1 330. They recognized four biogeographical regions or marine provinces, notably the Namaqua Province (southern Namibia to Cape Point, west of False Bay), the Algoa Province (Cape Agulhas to the Great Kei river mouth), the Natal Province (eastern Transkei and southern Natal to the Tugela River) and the Indo-Pacific Province (Tugela River northwards and across the tropical Indian and Pacific oceans). They also recognized three transitional zones, notably the Namaqua-West African overlap (northern Namibia), the Namaqua-Algoa overlap (False Bay to Cape Agulhas) and the Algoa-Natal overlap (western Transkei). These points are shown in Figure 1.1.

The known littoral South African mollusc fauna is dominated by Cape endemics and Indo-Pacific species, these two components together comprising roughly 80% to 90% of the total fauna. The analysis by Kilburn and Rippey (1982) shows a marked transition from Cape endemic to Indo-Pacific dominance between the Cape West Coast and Natal; this is shown in Table 8.2.

Table 8.2 Transition in the mollusc fauna from Cape endemic to Indo-Pacific dominance

	Cape Endemics (%)	Indo-Pacific (%)
Cape West Coast	88,2	0
False Bay	88,9	2,4
Cape Agulhas	88,4	5,0
Eastern Cape	70,5	16
Western Transkei	52,7	29
Eastern Transkei	38,4	40
Natal	18,7	62

This transition is clearly linked to the increasing influence of the warm Agulhas Current northwards up the east coast. An interesting phenomenon noted by Kilburn and Rippey (1982) is the reappearance in southern Natal and Transkei of many tropical Indo-Pacific molluscs. They attributed this to the frequent occurrence of local inshore eddies of tropical Agulhas Current water and to the presence of ecological niches which may be available for occupation in the absence of local competition.

Although the molluscs occupying the nearshore coastal zone of Natal are fairly well known, those from deeper water, particularly north of the Tugela, have yet to be fully investigated. Commercial trawlers have yielded some of the larger molluscs but many smaller species have been overlooked. Recent collections on the east coast continental slope have yielded a very high percentage of undescribed species (Kilburn, pers. comm.). Considerable effort is evidently still required to fully characterize the mollusc fauna of Natal.

Polychaeta

Marine polychaete worms are common, diverse and ubiquitous. They often dominate the marine benthos both in numbers of species and in numbers of specimens. The importance of polychaetes in marine and estuarine ecosystems was recognized at an early stage by J.H. Day of the University of Cape Town, and he has devoted much of his life to studying them. This led to the publication of a monograph on southern African polychaetes (Day, 1967) in which about 750 species are described. More recently Hartmann-Schröder and Hartmann (1974) made a systematic study of intertidal polychaetes in Angola, Namibia, South Africa and Mozambique, describing in the process one new genus, twenty new species and seven sub-species.

Day (1967) acknowledges that much work remains to be done, particularly on the shelf fauna of Natal and Mozambique, but has nevertheless made a preliminary zoogeographical analysis of the polychaete fauna from the intertidal zone to 200 m. For this he selected at random 100 species with good distribution records and divided them into six faunistic components on the basis of their world-wide distribution. These were Cosmopolitan, Circumtropical, Tropical Indo-West Pacific, Atlantic, Southern and Endemic. A seventh group, termed "Other Foreign" was erected to contain ill-defined non-endemic species. He then considered the distribution of these components around southern Africa. The percentage distributions at the main collection stations (see Figure 1.1) of the Endemic, Indo-Pacific and Circumtropical components, which are of particular interest to Natal are shown in Table 8.3.

The endemic component was particularly strong and showed a fairly constant presence along the west and south coasts. From the Mbashe river north-eastwards it declined abruptly. Conversely, the Indo-Pacific and Circumtropical components showed a low presence on the west and south coasts but increased in abundance from the Bashee river north-eastward. This increase is again a clear manifestation of the southward influence of the Agulhas current. In discussing the Indo-Pacific component, Day (1967) noted that it extended up the west coast into the cold Benguela current and suggested the possibility that the South African endemics have been derived from cold-tolerant fauna of Indo-West Pacific origin.

EFFECTS OF POLLUTION ON MARINE BENTHOS

The value of monitoring changes in marine benthic communities to assess the impact of pollution has long been recognized. Most benthic

Table 8.3 Percentage distributions of Polychaete fauna at the given collection stations

	Endemic (%)	Indo-pacific (%)	Circumtropical (%)
West Coast			
Walvis Bay) Namibia	55	0	0
Luderitz)	50	6	0
Port Nolloth	47	8	0
Lamberts Bay	47	6	0
South Coast			
Cape Point(False Bay)	42	5	5
Cape Agulhas	40	6	7
Knysna	42	5	8
Port Elizabeth	49	5	9
East Coast			
Mbashe River	32	12	16
Durban	19	26	23
Maputo	5	37	32
Inhambane	0	43	37

organisms have limited mobility and cannot simply move away from adverse conditions. Sensitive species may be eliminated whereas others may thrive. The net result is a change in community structure which may reflect the nature and severity of the impact.

In 1974 a national marine pollution survey was initiated with the object of determining and assessing "sources, concentration levels, pathways and consequence of pollutants in impact areas, in estuaries and at coastal reference stations around the coast of South Africa" (Cloete and Oliff, 1976). Part of the survey effort was devoted to Natal under the auspices of the National Institute for Water Research (NIWR) of the CSIR. The NIWR Natal laboratory had already been concerned with marine pollution since the early 1960s under contracts with local authorities and industries wishing to dispose of effluents to sea. Studies on the marine benthos have been an integral part of the NIWR surveys from the outset. With the implementation of the national marine pollution survey in 1974 the opportunity was taken to extend these surveys from specific polluted localities to include the whole Natal coastline from Kosi Bay to Port Edward.

Transects, each consisting of three stations where the water depths were 20, 50 and 100 m, were established at thirteen positions between Kosi Bay and Port Edward (Figure 8.1). A central deep-water transect was also established off Durban with stations at positions 2, 4, 6, 8, 10, 25, 50, 75 and 100 km offshore. The transects were visited on a number of occasions over a period of five years. In addition to a variety of physical and chemical measurements, and in order to establish a pollution baseline, surveys were made of the benthos. These data have been included in a marine pollution data base which is being administered by the South African Data Centre for Oceanography at the National Research Institute for Oceanology in Stellenbosch. They form the basis of on-going studies on the effects of pollution on marine benthos. It is anticipated that a better understanding of the marine benthos will gradually emerge from this research, particularly at Richards Bay, Durban, Umbogintwini and Umkomaas where significant quantities of effluent are being discharged to sea through offshore pipelines (see Chapter 10, for detailed descriptions of these pipelines).

Off Durban, two large submarine pipelines have been discharging municipal and industrial wastes at a water depth of about 50 m since the late 1960s. Surveys of the benthos made before and after introduction of the pipelines (National Institute for Water Research, 1968; 1972; 1974; 1983; 1984; 1985) have revealed that the discharges have had very little environmental impact on the sea bed. This is attributed to the dynamic current system which predominates in the area and the consequent rapid dilution and dispersion of effluent.

At Richards Bay a similar effluent disposal system, discharging at about 30 m water depth, was commissioned in 1985 (Chapter 11). Surveys of the benthos made there between 1972 and 1984 (McClurg *et al*. 1985) showed the presence of an impoverished bottom fauna over much of the area. This was evidently a result of extensive dumping of sediments dredged during harbour construction.

GENERAL COMPOSITION OF BENTHOS

Some idea of the general composition of the marine benthos off Natal can be gained from data obtained in the baseline and pollution monitoring surveys. Details of the benthos at 563 stations were available at the time of writing. Of these stations, 425 were sampled by cone dredge, 63 by Smith-McIntyre grab, 46 by Van Veen grab, 18 by beam trawl and 12 by heavy biological dredge. Only the cone dredge results were sufficiently detailed to warrant further analysis.

These samples, each of which comprised the fauna contained in one litre of sediment and retained by a 235 micron sieve, yielded 115 655 specimens representing 372 identifiable taxa. The composition of this fauna is reflected in the accompanying pie and bar charts (Figure 8.2). Nematode worms comprised numerically by far the greatest part of the fauna (Figure 8.2A) followed by annelids and arthropods. Since nematodes, turbellarians and harpacticoid copepods are traditionally regarded as meiofauna they may be considered separately. Their exclusion from the analysis (Figure 8.2B) emphasizes the dominance of the macrofauna by annelids and arthropods and shows the echinoderms and mollusc to have had a relatively small but significant presence. The annelids in turn comprised very largely polychaete worms (Figure 8.2C), representing 37 families of which the syllidae and spionidae were the most abundant (Figure 8.2E). The arthropods were made up almost entirely of crustaceans (99.4%) of which the amphipoda were predominant, comprising 48% of the total (Figure 8.2D). The most numerous of the 20 amphipod families were the Haustoriidae, Ampeliscidae, Corophiidae and Phoxocephalidae (Figure 8.2F).

Data collected on the deep-water transect off Durban (Figure 8.1) give some idea of the changes in benthic diversity and abundance with water depth and distance offshore. Eleven surveys extending to 25 km offshore and a water depth of about 450 m were available for analysis. To allow intercomparison between surveys the counts were normalized by expressing them as percentages of the total. The faunal counts and numbers of species both showed a marked decline inshore and offshore of station 2 (Figure 8.3). Several physical factors varying with increasing depth can be invoked to account for this distribution pattern. These include decrease in wave-induced turbulence, lower light intensity and temperature, and change in sediment type. It is interesting to note the relationship that exists between the benthos distribution, described here, and the three sedimentary zones parallel to the shore recognized by Flemming (1980) on the east coast. Station 2, where the greatest abundance of benthos was found, falls into the

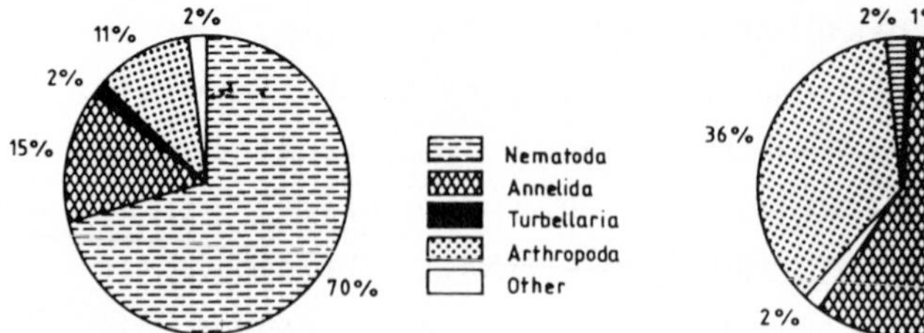

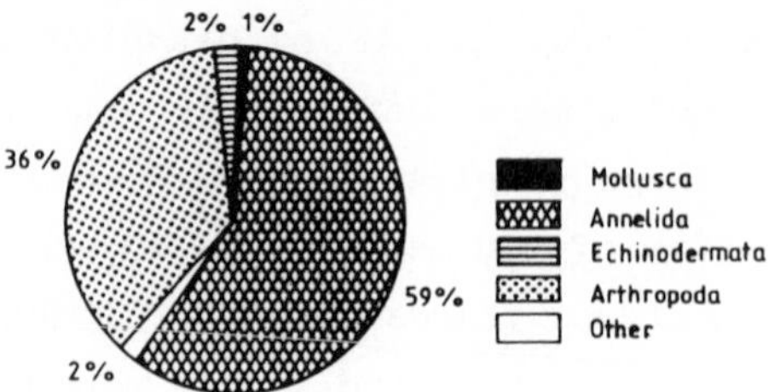

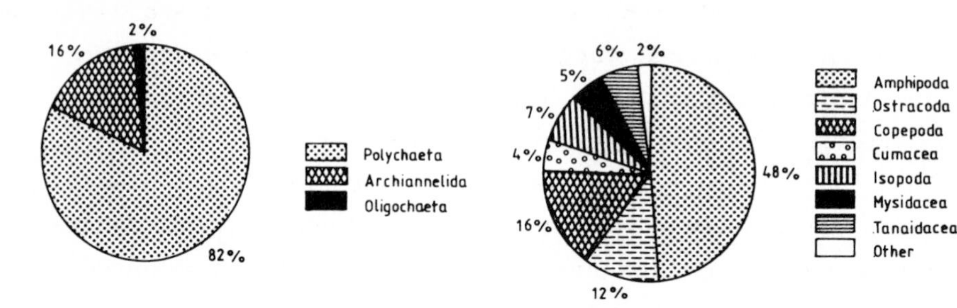

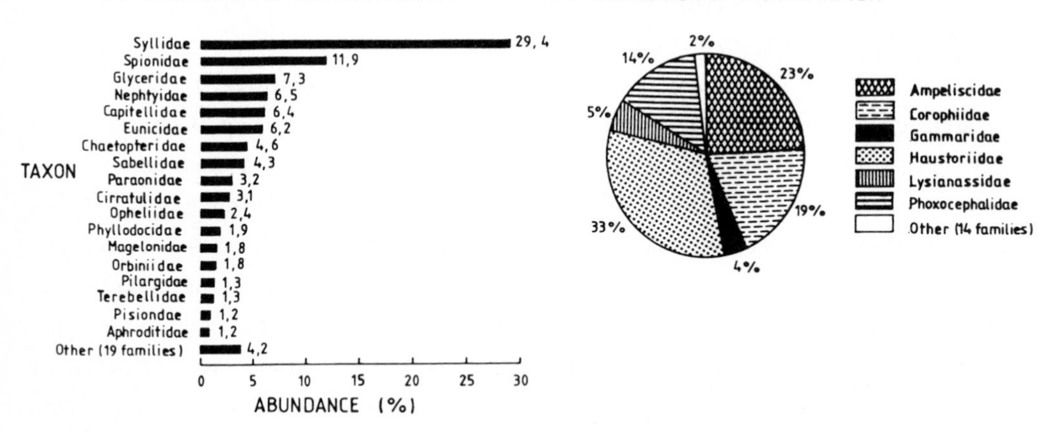

Figure 8.2 Pie and bar charts showing the composition of benthos sampled by cone dredge at 425 stations on the Natal shelf.

zone described by Flemming as the "current controlled central-shelf sand stream". The inshore station 1 falls within the "wave-dominated nearshore sediment wedge", whereas stations 3 and 4, further offshore,

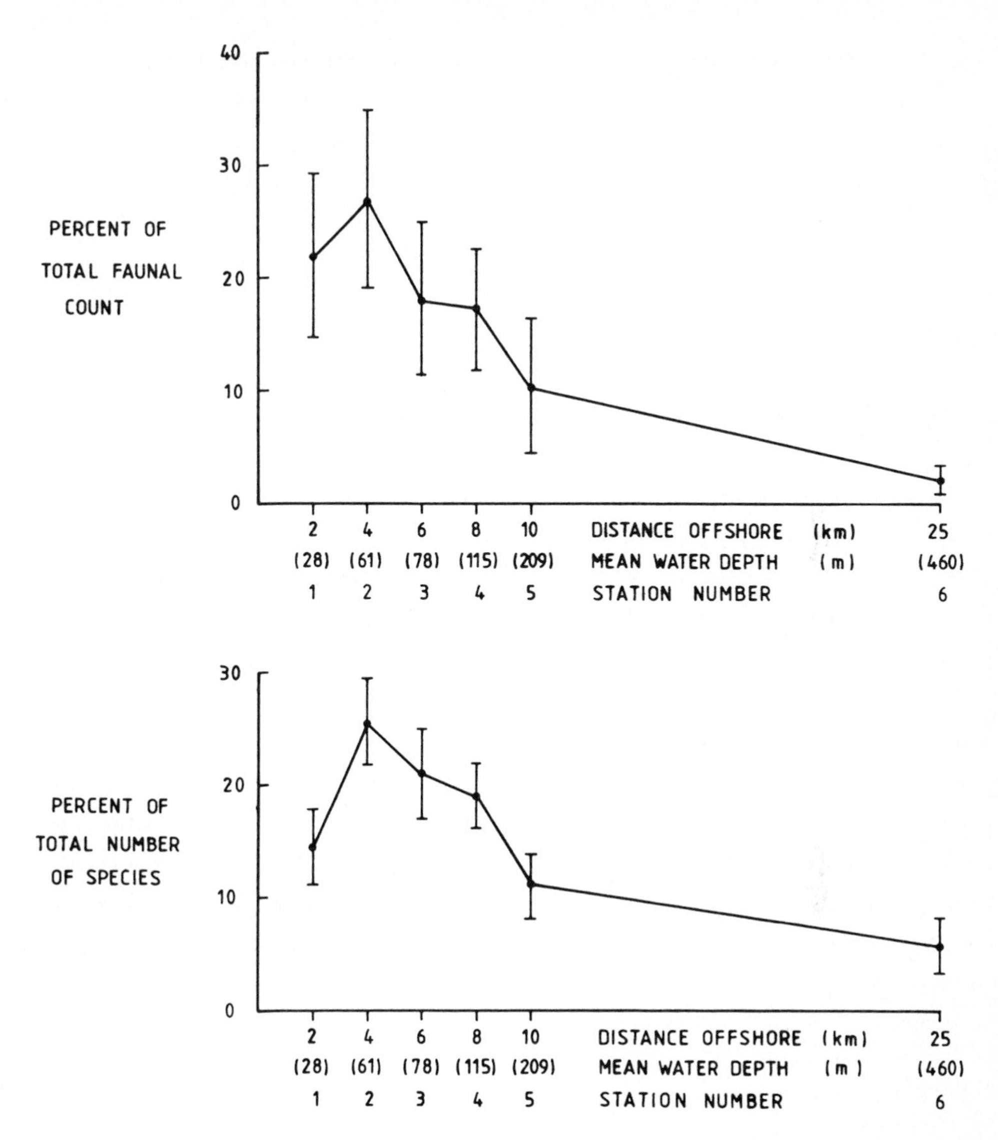

Figure 8.3 Variations in the abundance and diversity of benthos with distance offshore on the Durban transect. The mean value and 95% confidence limits for eleven surveys are shown for each station.

are situated on the "sand depleted outer-shelf gravel pavement". Stations 5 and 6 are in deeper water beyond the shelf break. It thus appears that "central-shelf sand stream" zone provides optimal conditions for the development of benthic communities off Durban. An attempt to relate the benthos distribution to sediment characteristics (organic content and grain size) proved futile because of wide variations between the inadequately low number of samples. A more detailed study will be required before a definitive statement can be made regarding animal/sediment relations on the Natal continental shelf.

GENERAL BIOGEOGRAPHICAL STUDIES

Although there have been no specific biogeographical accounts of the Natal marine fauna, the region has been included in many major biogeographical texts (e.g. Forbes 1856; Ekman 1953; Stephenson and Stephenson 1972; Briggs 1974; Brown and Jarman 1978). The latter two authors provide a useful description of coastal marine habitats around southern Africa and review much of the earlier thinking on biogeographical provinces. They remark that present concepts of biogeographical division depend largely on the mode of the life of the species being considered, with a slightly different pattern emerging for each group of organisms. However, ultimately it is the broad trends that are important and these clearly show the Natal marine fauna to fall within a distinct east coast subtropical province extending from southern Mozambique to the Transkei. The fauna derives its basic character from the tropical Indo-West Pacific and receives a continuing influence from this region through the southward-flowing warm Agulhas current.

Acknowledgements

This contribution is published with the permission of the Director of the National Institute for Water Research, while financial support of the Foundation for Research Development of the CSIR is also gratefully acknowledged. Drs A D Connell, A J de Freitas and R N Kilburn gave useful advice and comments on the manuscript, while Mr R Singh was responsible for reproducing the figures.

REFERENCES

ANONYMOUS (1984). South African Asociation for Marine Biological Research. Oceanographic Research Institute. Report on research during the period April 1983 to March 1984. Durban, March 1984, 42 pp.

BARNARD, K H (1924a). Contributions to the crustacean fauna of South Africa. No. 7 - Cirripedia. **Annals of the South African Museum,** 21(1), 1-103.

BARNARD, K H (1924b). Descriptions of a new genus and species of isopod crustacean belonging to the family Bathynomidae, procured in the South African Marine Survey. **Report Fisheries and Marine Biological Survey, Union of South Africa.** 4(1923). Special Report 11, 1-4.

BARNARD, K H (1940). Contributions to the crustacean fauna of South Africa. XII. Further additions to the Tanaidacea, Isopoda and Amphipoda, together with keys for the identification of hitherto recorded marine freshwater species. **Annals of the South African Museum,** 32(5), 293-543.

BARNARD, K H (1950). Descriptive catalogue of South African decapod crustacea (crabs and shrimps). **Annals of the South African Museum,** 38, 1-837.

BARNARD, K H (1954). South African Pyncogonida. **Annals of the South African Museum,** 39(3), 81-159.

BARNARD, K H (1957). Three additions to the fauna list of South African Crustacea. **Annals and Magazine of Natural History,** (12) 10, 814-816.

BARNARD, K H (1964). The work of the S S Pieter Faure in Natal waters, with special reference to the Crustacea and Mollusca; with descriptions of new species of Mollusca from Natal. **Annals of the Natal Museum,** 16, 9-29.

BERRY, P F (1969). The biology of Nephrops andamanicus Wood-Mason (Decapoda, Reptantia). **Investigational Report of the Oceanographic Research Institute,** 22, 1-61.

BERRY, P F (1973). The biology of the spiny lobster Palinurus delagoae Barnard, off the coast of Natal, South Africa. **Investigational Report of the Oceanographic Research Institute,** 31, 1-27.

BERRY, P F and R PLANTE (1973). Revision of the spiny lobster genus Palinurus in the south-west Indian Ocean. **Transactions of the Royal Society of South Africa,** 40, 373-380.

BERRY, P F, A E F HEYDORN and D J ALLETSON (1975). The biology of the knife prawn Hymenopenaeus triarthrus off the Natal coast. Durban Oceanographic Research Institute, 23 pp. (Unpublished report).

BISESWAR, R (1985). The geographic distribution of Echiura from southern Africa. **South African Journal of Marine Science,** 3. 11-21.

BOROJEVIC, R (1967). Spongaires d'Afrique du Sud. **Transactions of the Royal Society of South Africa.** 37(3), 183-226.

BOSHOFF, P H (1980). The corals of Maputaland. In: **Studies on the ecology of Maputaland.** (Eds: M N BRUTON and K H COOPER). Rhodes University and the Natal Branch of the Wildlife Society of Southern Africa, 111-113.

BOSHOFF, P H (1981). An annotated checklist of southern African Scleratina. **Investigational Report of the Oceanographic Research Institute,** 49, 1-45.

BRIGGS, J C (1974). **Marine zoogeography.** McGraw Hill, New York.

BROWN, A C (1956). Additions to the genus Apseudes (Crustacea:Tanaidacea) from South Africa. **Annals and Magazine of Natural History** (ser. 12) 9, 705.

BROWN, A C (1957). A revision of the genus Leptochelia (Crustacea:Tanaidacea) in southern African waters. **Annals and Magazine of Natural History** (ser. 12) 10, 401.

BROWN, A C and N JARMAN (1978). Coastal marine habitats. In: **Biogeography and ecology of southern Africa.** (Ed: M S A Werger). Volume II, 1241-1277.

CARLGREN, O (1938). South African Actinaria and Zoantharia. **Kungliga Svenska vetenskapsakademiens handlinger** (ser. 3) 17(3), 1-148.

CHAMPION, H F B (1970). Aspects of the biology of Penaeus indicus Milne-Edwards with notes on associated Penaeidae occurring off Natal on the east coast of South Africa. Oceanography in South Africa - 1970. Durban. 4-6 August, 1970 (G1): 1-17 (Unpublished paper).

CHAMPION, H F B (1976). Recent prawn research at St Lucia. Scientific Advisory Workshop Meeting at Charters Creek, 15-17 February, 1976. Natal Parks, Game and Fish Preservation Board, Pietermaritzburg.

CHERBONNIER, G (1952). Contribution á la connaissance des holothuries de L'Afrique du Sud. **Transactions of the Royal Society of South Africa,** 33, 469-509.

CHERBONNIER, G (1970). Nouvelles espéces d'Holothuries des côtes d'Afrique du Sud et du Mozambique. **Bulletin de Muséum d'histoire naturelle. Paris,** 41(1), 280-299.

CLARK, A M (1977). The South African Museum's R.V. Meiring Naudé cruises, Part 4: Echinoderms. **Annals of the South African Museum,** 73(6), 133-147.

CLARK, A M and J COURTMAN-STOCK (1976). **The echinoderms of southern Africa.** British Museum (Natural History), London, 277 pp.

CLARK, H L (1923). The echinoderm fauna of South Africa. **Annals of the South African Museum,** 13(7), 221-435.

CLOETE, C E and W D OLIFF (1976). South African marine pollution survey report 1974-1975. **South African National Scientific Programmes Report 8.** 60pp.

COOPER, G A (1973a). New Brachiopoda from the Indian Ocean. **Smithsonian Contributions to Paleobiology,** 16, 1-43.

COOPER, G A (1973b). Vema's Brachiopoda (Recent). **Smithsonian Contributions to Paleobiology,** 17, 1-51.

COOPER, G A (1981). Brachiopoda from the southern Indian Ocean (Recent). **Smithsonian Contributions to Paleobiology,** 43, 1-93.

CROSSLAND, C (1948). Reef corals of the South African coast. **Annals of the Natal Museum,** 11(2), 169-205.

DAY, J A (1975). South African Cumacea, Part 1: Family Bodotriidae, Subfamily Vaunthompsoniinae. **Annals of the South African Museum,** 66(9), 177-220.

DAY, J A (1978a). Southern African Cumacea, Part 2: Family Bodotriidae, Subfamily Bodotriinae. **Annals of the South African Museum,** 75(7), 159-290.

DAY, J A (1978b). South African Cumacea, Part 3: Families Lampropidae and Ceratocumatidae. **Annals of the South African Museum,** 76(3), 137-189.

DAY, J A (1980). Southern African Cumacea, Part 4: Families Gynodiastylidae and Diastylidae. **Annals of the South African Museum,** 82(6), 187-192.

DAY, J H (1967). **A monograph on the polychaeta of southern Africa.** Trustees of the British Museum (Natural History), London. 878 pp.

DE FREITAS, A J (1980). **The Penaeoidea of south east Africa.** PhD Thesis, University of the Witwatersrand, Johannesburg, 480 pp.

DE FREITAS, A J (1984a). The Penaeoidea of south east Africa. I. The study area and key to the south east African species. **Investigational Report of the Oceanographic Research Institute,** 56.

DE FREITAS, A J (1984b). The Penaeoidea of south east Africa. V. The family Sicyoniidae. **Investigational Report of the Oceanographic Research Institute,** 60.

DE FREITAS, A J (1985a). The Penaeoidea of south east Africa. II. Families Aristeidae and Solenoceridae. **Investigational Report of**

the Oceanographic Research Institute, 57.

DE FREITAS, A J (1985b). The Penaeoidea of south east Africa. III. Family Penaeidae (excluding genus Penaeus). **Investigational Report of the Oceanographic Research Institute,** 58.

DE FREITAS, A J (1985c). The Penaeoidea of south east Africa. IV. Family Penaeidae, genus Penaeus. Investigational **Report of the Oceanographic Research Institute,** 59.

DEICHMANN, E (1948). The holothurian fauna of South Africa. Annals of the Natal Museum, 11, 325-376.

EKMAN, S (1953). **Zoogeography of the sea.** Sidwick and Jackson, London.

FLEMMING, B W (1980). Sand transport and bedform patterns on the continental shelf between Durban and Port Elizabeth (south east African continental margin). **Sedimentary Geology,** 26, 179-205.

FLEMMING, B W (1981). Factors controlling shelf sediment dispersal along the south east African continental margin. **Marine Geology,** 42, 259-277.

FORBES, A T and M C BENFIELD (1985). Aspects of the penaeid prawn fisheries in Natal. **South African Journal of Science,** 81, 430-431.

FORBES, E (1856). Map of the distribution of marine life. In: **The physical atlas of natural phenomena.** (Ed: A K JOHNSTON). W and A K Johnston, Edinburgh.

GRIFFITHS, C L (1974). The Amphipoda of southern Africa, Part 3: The gammaridea and Caprellidea of Natal. **Annals of the South African Museum,** 62(7), 209-264.

GRIFFITHS, C L (1976). Guide to the benthic marine amphipods of southern Africa. Trustees of the South African Museum, Cape Town. 106 pp.

GRIFFITHS, C L (1977). The South African Museum's Meiring Naudé cruises, Part 6: Amphipoda. **Annals of the South African Museum,** 74(4), 105-123.

HARTMANN-SCHRODER, G and G HARTMANN (1974). Zur Kenntnis des Eulitorals der afrikanischen Westküste zwischen Angola und Kap der Guten Hoffnung und der afrikanischen Ostküste van Südafrika und Mozambique unter besonderer Berücksichtigung der Polychaeten und Ostracoden. **Mitteilungen aus dem Hamburgischer zoologischen Museum und Institut,** 69, 1-514.

HAYWARD, P J and P L COOK (1979). The South African Museum's Meiring Naudé cruises, Part 9: Bryozoa. **Annals of the South African Museum,** 79(4), 43-130.

HAYWARD, P J and P L COOK (1983). The South African Museum's Meiring Naudè cruises, Part 13, Bryozoa II. **Annals of the South African Museum,** 91(1), 1-161.

HEYDORN, A E G, N D BANG, A F PEARCE, B W FLEMMING, R A CARTER, M H SCHLEYER, P F BERRY, G R HUGHES, A J BASS, J H WALLACE, R P VAN DER ELST, R J M CRAWFORD and P A SHELTON (1976). Ecology of the Agulhas current region: an assessment of biological responses to environmental parameters in the south west Indian Ocean. **Transactions of the Royal Society of South Africa,** 43(2), 151-190.

HILLER, N (1986). The South African Museum's Meiring Naudé cruises, Brachiopoda from the 1975-1979 cruises. **Annals of the South African Museum,** (in press).

HULLEY, P A (1984). The South African Museum's Meiring Naudé cruises, Part 14: Family Myctophidae (Osteichthyes, Myctophiformes) **Annals of the South African Museum,** 93(2), 53-96.

JACKSON, J W (1952). A revision of some South African Brachiopoda; with descriptions of new species. **Annals of the South African Museum,** 41(1), 1-40.

JOUBERT, L S and D H DAVIES (1966). The penaeid prawns of the St Lucia Lake system. **Investigational Report of the Oceanographic Institute,** 13, 1-40.

KENSLEY, B F (1972). Shrimps and prawns of southern Africa. Trustees of the South African Museum, Cape Town. 65pp.

KENSLEY, B F (1977a). The South African Museum's Meiring Naudé cruises, Part 2: Crustacea, Decapoda, Anomura and Brachyura. **Annals of the South African Museum,** 72(9), 161-188.

KENSLEY, B F (1977b). The South African Museum's Meiring Naudè cruises, Part 5: Crustacea, Decapoda, Reptantia and Natantia. **Annals of the South African Museum,** 74(2), 13-44.

KENSLEY, B F (1978a). The South African Museum's Meiring Naudè cruises, Part 7: Marine Isopoda. **Annals of the South African Museum,** 74(5), 125-157.

KENSLEY, B F (1978d). A new marine isopod family from the south-western Indian Ocean. **Annals of the South African Museum,** 75, 41-50.

KENSLEY, B F (1979). A second genus of the marine isopod family Bathynataliidae. **Annals of the South African Museum,** 79(3), 35-41.

KENSLEY, B F (1981a). The South African Museum's Meiring Naudè cruises, Part 12: Crustacea, Decapoda of the 1977, 1978, 1979 cruises. **Annals of the South African Museum,** 83(4), 49-78.

KENSLEY, B F (1981b). On the zoogeography of southern African decapod crustacea, with a distributional checklist of the species. **Smithsonian Contributions to Zoology,** Number 338. Smithsonian Institution Press, Washington, 64 pp.

KENSLEY, B F (1982). Revision of the southern African Anthuridea (Crustacea, Isopoda) **Annals of the South African Museum,** 90(3), 95-200.

KENSLEY, B F (1984). The South African Museum's Meiring Naudè cruises, Part 15: Marine Isopoda of the 1977, 1978, 1979 cruises. **Annals of the South African Museum,** 93(4), 213-301.

KILBURN, R J and E RIPPEY (1982). Sea shells of southern Africa. Macmillan South Africa (Publishers) (Pty) Ltd. Johannesburg.

LEVI, C (1963). Spongaires d'Afrique du Sud. (2): Poecilosclerides. **Transactions of the Royal Society of South Africa,** 37(1), 1-83.

LEVI, C (1967). Spongaire d'Afrique du Sud. (3): Tetractinellidae. **Transactions of the Royal Society of South Africa,** 73(3), 227-256.

LOUW, E (1977). The South African Museum's Meiring Naudè cruises, Part 1: Station data 1975, 1976. **Annals of the South African Museum,** 72(8), 147-159.

LOUW, E (1980). The South African Museum's Meiring Naudè cruises, Part 10: Station data 1977, 1978, 1979. **Annals of the South African Museum,** 81(5), 187-205.

MacNAE, W (1962). The fauna and flora of the eastern coasts of southern African in relation to ocean currents. **South African Journal of Science,** 58(7), 208-212.

McCLURG, T P (1984). Effects of fluoride, cadmium and mercury on the estuarine prawn Penaeus indicus. **Water S.A.** 10(1), 40-45.

McCLURG T P, B D GARDNER and N S PAYNTER (1985). Benthic Macrofauna. In: Connell A D, T P McClurg and D J Livingstone (eds.). Environmental studies at Richards Bay prior to the discharge of submarine outfalls: 1974-1984. Marine Research Group, National Institute for Water Research, Durban Branch Laboratory, pp 45-96.

MILLARD, N A H (1958). Hydrozoa from the coasts of Natal and Portuguese East Africa. Part I. Calyptoblastea. **Annals of the South African Museum,** 44, 165-226.

MILLARD, N A H (1959). Hydrozoa from the coasts of Natal and Portuguese East Africa. Part I. Gymnoblastea. **Annals of the South African Museum,** 44, 297-313.

MILLARD, N A H (1975). Monograph on the Hydroida of Southern Africa. **Annals of the South African Museum,** 68, 1-513.

MILLARD, N A H (1977). The South African Museum's Meiring Naudè cruises, Part 3: Hydroida. **Annals of the South African Museum,**

73(5), 105-131.

MILLARD, N A H (1978). The geographical distribution of southern African hydroids. **Annals of the South African Museum,** 82 (4), 129-153.

NATIONAL INSTITUTE FOR WATER RESEARCH (1968). A survey of the bacteriology, chemistry and biology of the sea and beaches in the vicinity of Durban. Conditions prior to the use of submarine outfalls. Period 1964-1968. CSIR Contract Report C Wat 19, Pretoria.

NATIONAL INSTITUTE FOR WATER RESEARCH (1974). Surveys monitoring the sea and beaches in the vicinity of Durban. Part 2. Conditions following the use of the submarine outfalls CSIR Contract Report C Wat 26. Durban.

NATIONAL INSTITUTE FOR WATER RESEARCH (1974). Surveys monitoring the sea and beaches in the vicinity of Durban. Part 3. Surveys between 1972 and 1974. CSIR Contract Report C Wat 30. Durban.

NATIONAL INSTITUTE FOR WATER RESEARCH (1983). Sea disposal of sludge: detailed report on environmental surveys in the Durban outfall region, with technical appendices. CSIR Contract Report C Wat 55. Durban.

NATIONAL INSTITUTE FOR WATER RESEARCH (1984). Sea disposal of sludge' environmental surveys in the Durban outfalls region. Part 2. Surveys made between June 1983 and May 1984. CSIR Contract Report C Wat 58. Durban.

NATIONAL INSTITUTE FOR WATER RESEARCH (1985). Sea disposal of sewage: environmental surveys in the Durban outfalls region. Part 3. Surveys made between June 1984 and May 1985. CSIR Contract Report C Wat 67. Durban.

O'DONOGHUE, C H (1957). Some South African Bryozoa. **Transactions of the Royal Society of South Africa,** 35(2), 71-95.

O'DONOGHUE, C H and D de WATTEVILLE (1937). Notes on South African Bryozoa. **Zoologischer Anzeiger,** 117, 12-22.

O'DONOGHUE, C H and D de WATTEVILLE (1944). Additional notes on Bryozoa from South Africa. **Annals of the Natal Museum,** 10(3), 407-432.

STEBBING, T R R (1910). Sympoda. **Annals of the South African Museum,** 6, 409-419.

STEBBING, T R R (1912). South African Crustacea, Part 6. The Sympoda. **Annals of the South African Museum,** 10, 129-176.

STEPHEN, A C (1942). The South African intertidal zone and its relation to ocean currents. Notes on the intertidal sipunculids of Cape Province and Natal. **Annals of the Natal Museum,** 10(2),

245-256.

STEPHEN, A C and E B CUTLER (1969). On a collection of Sipuncula, Echiura and Priapulida from South African waters. **Transactions of the Royal Society of South Africa,** 38(2), 111-121.

STEPHENSON, T A and A STEPHENSON (1972). **Life between tidemarks on rocky shores.** W H Freeman and Company, San Francisco. 425 pp.

THANDAR, A S (1971). **The intertidal holothurian fauna of the rocky shores of Natal.** MSc thesis, University of South Africa. 235 pp.

THANDAR, A S (1977). Description of two new species of Holothuroidea from the east coast of South Africa. **Annals of the Natal Museum.** 23(1), 57-66.

THANDAR, A S (1984). **The Holothurian fauna of southern Africa.** PhD thesis, University of Durban-Westville. 566 pp.

THANDAR, A S (1985). A new southern African genus in the holothurian family Cucumariidae (Echinodermata: Holothuroidea) with the recognition of two subspecies in Cucumaria frauenfeldi Ludwig. **South African Journal of Zoology,** 20(3), 109-114.

TIXIER-DURIVAULT, A (1954). Lec octocoralliares d'Afrique du Sud. I. Alcyonacea, II Gorgonacea, III Pennatulacea. **Bulletin du Muséum d'histoire naturelle. Paris.** (ser. 2) 22, 124-129; 261-268; 385-390; 526-533; 624-631.

WEBB, M (1972). Ochaetostoma erthrogrammon (Leuckart and Ruppell 1828) (Echiurida) from Isipingo beach, Natal, South Africa. **Zoologica Africana,** 7(2), 521-532.

WESENBERG-LUND, E (1963). South African sipunculids and echiuroids from coastal waters. **Videnskabelige Meddelelser fra Dansk naturhistorisk Forening i Kjobenhaun** 126, 101-146.

WHEELER, J F (1940). Some nemerteans from South Africa and a note on Lineus corrugatus McIntosh. **Journal of the Linnean Society (Zoology)** 42, 20-49.

Chapter 9

SHELF ICHTHYOFAUNA OF NATAL

Rudy van der Elst
Oceanographic Research Institute, Durban

ZOOGEOGRAPHY

Fishes are the most numerous vertebrates on earth, in terms of both abundance and number of species. Of approximately 40 000 backboned animals that have been described, about 19 000 are fish, and 64% of these are marine or estuarine (Nelson, 1976). No other group of vertebrates can match the diversity in form and habitat preference of fishes, and no naturally occurring group of 'wild' animals provides as much protein and wealth to man. Fish have successfully invaded virtually all waters at the known extremes of temperature and salinity, although physico-chemical factors obviously influence their dispersal. Temperature is generally considered to be the single most important determinant in fish zoogeography (Cohen, 1973) and Hubbs (1952) has suggested how changes in Pleistocene water temperatures could have influenced the present distribution of marine fishes. Most striking is the 'antitropical' distribution of fish, where similar families, genera or species occur to the north and south of the tropics but not within them. Cooler waters during the ice ages (Berg, 1933) or deep-water cold isotherms that crossed the Pleistocene tropics (Hubbs, 1952) are probable explanations for 'antitropical' distributions.

Although marine ichthyofauna is widely dispersed throughout the world, regions can be identified on the basis of distinct faunal composition (Myers, 1941; Ekman, 1953; MacArthur and Wilson, 1967; Cohen, 1973). A total of eight such regions have been documented, each with a distinct presence of endemic forms: Indo-Pacific (Indian and West Pacific), tropical east Atlantic, tropical west Atlantic (includes tropical east Pacific), north Pacific, north Atlantic, Mediterranean-east Atlantic, Arctic and Antarctic (Ekman, 1953; Nelson, 1973). There is little doubt that the Indo-Pacific region is richest of all as it "... contains practically all the families and a considerable number of genera that make up the fauna of the other three (tropical) zones" (Myers, 1941). Much evidence also exists to suggest that the Indo-Pacific marks the central point for marine fish speciation, which led Hubbs (1952) to term it "the great mother fish fauna of the world".

The Indian Ocean itself does not appear to be a natural zoogeographic unit, as very few species which live in both the western and eastern parts do not live in the Pacific Ocean (Cohen, 1973). Nevertheless, there are several areas of endemism within the Indian Ocean, i.e. the Red Sea, the Arabian Sea and the southwestern Indian Ocean (Cohen, op. cit.). The Red Sea is more obviously distinct in that its narrow and shallow entrance makes it geographically more isolated and physically distinguishable in terms of temperature and salinity. Consequently its ichthyofauna is rich in endemics and apparently with speciation tendencies (Steinitz, 1973). The Arabian Sea and Persian Gulf are somewhat less well defined but nevertheless their ichthyofauna also tends to be partially localized (Nelson, 1973). Geographically less distinct is the southwestern Indian Ocean, yet evidence exists that there, too, the ichthyofauna has characteristics that distinguish it from that in the rest of the Indian Ocean (Nelson, 1973).

Reasons for isolating the southwestern Indian Ocean fauna (which includes the Natal fauna) can be further discussed. Sea surface temperatures recorded during the International Indian Ocean Expedition (Wyrtki, 1971) clearly demonstrate that both Natal and Western Australia lie predominantly between the 20°C and 25°C isotherms with spells of water temperatures lower than 20°C off Natal occurring with some frequency (see Fig. 5.4). It is well known that 20°C represents the limit of reef-building corals (Ekman, 1953) and it must be assumed, therefore, that temperatures below this occur often enough to limit reef formation. As coral reefs support a prolific and characteristic ichthyofauna, this temperature limit has a pronounced effect on fish zoogeography and clearly contributes to maintaining the Natal ichthyofauna distinct. Similarly the surface salinity regimes off Natal and Western Australia generally exceed 35×10^{-3}, whereas most of the remaining Indian Ocean is lower (Wyrtki, 1971); this could contribute to the characteristics of Natal's ichthyofauna.

The physical distinction of the South West Indian Ocean is quite pronounced at greater depths and the Wyrtki (1971) atlas reveals noticeably low concentrations of phosphates, nitrates and silicates to depths of 4 000 m. Not only are many of these physical features the same off Western Australia, but at depths exceeding 200 m there are quite a number of similarities between the waters of the Natal region and those on the east coast of Madagascar. The exact influence of these physical variables may not be understood, but they could well have some bearing on the characteristics of the ichthyofauna.

Ocean currents are known to be prime controllers of fish distribution, not only because they influence the temperature and chemical regimes of regions, but also because they disperse fish eggs and larvae. The larval stages of most tropical fish are extremely shortlived, hence many oceanic islands lack the ichthyofauna common to the mainland. For this reason the Hawaiian islands are known to lack rockcods and snappers (Oda and Parrish, 1981). Similarly, confined current systems will ensure that the juveniles of species are continuously recruited to the same region. It has been shown that circulation patterns on the Natal shelf indicate frequent current reversals (Chapter 5) and, despite the strong Agulhas Current, water masses and eddies often remain localised, further restricting the distribution of the ichthyofauna.

Indo-Pacific fishes number 6 000-7 000 species (Carcasson, 1977), the majority of which are also known from the Indian Ocean. Most are tropical shore fishes (3 000-4 000 according to Cohen, 1973), with 1 192 species belonging to 150 families having been reported from the Natal shelf region (Smith, 1980). Although Natal marine fishes are primarily Indo-Pacific in character, it is nevertheless possible to divide this fauna into five distinct categories based on the origin or distribution of their components. Hence after Smith (1980):

1)	Endemic	- 189	species	= 15.9%
2)	Indo-Pacific	- 879	"	= 73.8%
3)	Atlantic from West Africa	- 21	"	= 1.8%
4)	Circumglobal	- 100	"	= 8.4%
5)	Southern Ocean	- 3	"	= 0.3%

This incidence of endemism is high and compares rather surprisingly closely with 15% for the Red Sea and 14.4% for the Mediterranean

(Briggs, 1974). The commercially important endemic species belong to 27 families, though six families are particularly well represented, namely Cheilodactylidae = fingerfishes (4 out of 4); Clinidae = clingfishes (28 out of 36); Coracinidiae = galjoens (2 out of 2); Congiopodidae = horsefishes (2 out of 2); Kyphosidae = chubs (1 out of 2); Mugilidae = mullets (5 out of 15) and the Sparidae = seabreams (24 out of 32). The majority of these endemics tend to inhabit the more temperate regions of the South African east coast or are most abundant during the cooler months. Of further interest is the similarity between three of the families (Sparidae, Coracinidae and Kyphosidae) which are phylogenetically closely related (Greenwood et al. 1966), morphologically similar in appearance and tend to inhabit similar habitats (van der Elst, 1981).

It can thus be concluded that the Natal ichthyofauna is rich in species primarily of tropical Indo-Pacific origin. Nevertheless, the comparatively high incidence of endemism confirms that the south-west Indian Ocean, especially off Natal, has a characteristic ichthyofauna which can be distinguished from that elsewhere in the Indian Ocean.

NATURAL HISTORY OF NATAL FISHES

Obviously it is impossible to generalize on the life histories of members of an ichthyofauna as diverse as that found off Natal. Nonetheless, there are many unique and significant features which can be generally ascribed to the fishes of this region.

Foremost is their seasonality. Whereas a certain small proportion are present in Natal waters throughout the year, most are distinctly seasonal, being either summer or winter species.

The summer species are of two types: pelagic species that undertake long-distance migrations from tropical Indian Ocean waters, and species that appear to be more localized migrants from southern Mozambique. The former include such economically valued forms as the king mackerel (Scomberomorus commerson), queen mackerel (S. plurilineatus), rainbow runner (Elagatis bipinnulatus), wahoo (Acanthocybium solandri), yellowfin tuna (Thunnus albacares), skipjack (Katsuwonus pelamis), tenpounder (Elops machnata) and billfish (van der Elst, 1980, 1981).

Their population in Natal waters generally increases when the temperature of the surface water rises above 23°C (van der Elst and Collette, 1984), especially during the period December to May (Figure 5.4).

A feature common to many of these nomadic summer migrants is their reproductive inactivity. Although not all have been biologically investigated, those that have been studied fail to show any signs of gonad maturation, e.g. the queen mackerel (van der Elst and Collette, 1984), the tenpounder (Wallace, 1975b) and as yet unpublished records of king mackerel and billfish. The more localized summer migrants are mostly demersal or estuarine species and are frequently also caught, in limited numbers, during other seasons. Their reproductive activity is quite different and all are engaged in spawning runs during their period of abundance in Natal. Hence the spotted grunter (*Pomadasys commersonni*) and the perch (*Acanthopagrus berda*) (Wallace, 1975b), the slinger (*Chrysoblephus puniceus*) (Garratt, 1984) and the stonebream (*Neoscorpis lithophilus*) (Joubert, 1981) are summer spawners. It would appear that many of these summer spawners depend on the Natal littoral and estuarine environments for the recruitment of their juveniles, and many of their nursery areas have been identified (Wallace and van der Elst, 1975; Joubert, 1981; Berry *et al*. 1984).

Not only do tropical fish become more abundant during summer, but many other warm-water species are assisted by the Agulhas Current in penetrating further south, often reaching the eastern and southern Cape coasts. Examples include the cardinal-, damsel-, butterfly-, angel-, trigger- and parrotfishes (Smith, 1949; van der Elst, Crawford and Shelton, 1978). As many are attractive aquarium fishes, their seasonal appearance is eagerly awaited by amateur aquarists in the Cape.

In contrast to the wide-ranging summer fish, most of the winter species are either endemic or comprise isolated populations in South African waters, and all migrate from temperate Cape waters to Natal once temperatures drop below 21°C. This phenomenon of winter migrations to Natal often assumes massive proportions and may well be the most significant biological event that occurs annually in this ecosystem.

The onset of winter surface temperatures results in a whole different range of marine biota from an increase in the calanoid copepod population (Carter and Schleyer, 1978) to the appearance of the common dolphin _Delphinus delphis_ (Ross, 1984). Most 'dramatic' are the winter migrations of the pilchard _Sardinops ocellata_. Known colloquially as the Natal Sardine Run, shoals of these prime 'fodder' fish move close alongshore, often being stranded on shallow beaches; this is an event that is thought to result from nearshore circulation, onshore winds and, especially, attempts to avoid predators (Baird, 1971; van der Elst, Crawford and Shelton, 1978).

These predators are no less impressive in their sudden appearance and frequently form into frenzied feeding packs. Dominant amongst these are the copper shark (_Carcharhinus brachyurus_) (Bass, D'Aubrey and Kistnasamy, 1973), the elf (_Pomatomus saltatrix_) (van der Elst, 1976), leervis (_Lichia amia_) (van der Elst, 1981), the common dolphin (_Delphinus delphinus_) (Ross, 1984) and the Cape gannet (_Sula capensis_) (McLachlan and Liversidge, 1978), all being migrants from temperate Cape waters. The seasonal abundance of pilchards is quite clearly a feast for the predators and represents a significant input of nutrients to the Natal ecosystem.

Many other species are also known to migrate from the Cape to Natal at this time, though not all are such obvious predators of the pilchard. Included are the seventyfour (_Polysteganus undulosus_) (Ahrens, 1964), the red steenbras (_Petrus rupestris_), geelbek (_Atractoscion aequidens_), yellowtail (_Seriola lalandi_), poenskop (_Cymatoceps nasutus_)(van der Elst, 1981), the kob (_Argyrosomus hololepidotus_) (Wallace, 1975b) and several others. Whereas most are predators of the pilchard and no doubt synchronise their migrations accordingly, others are not. The karanteen (_Sarpa salpa_), for example, is a typical winter migrant from the Cape (Joubert, 1981) as is the Cape stumpnose (_Rhabdosargus holubi_)(Wallace, 1975a), yet both are herbivores. Of major significance is the fact that without exception all these winter migrants spawn in Natal. The benefit of such a common strategy is soon evident when the Agulhas Current is seen to serve as a dispersal mechanism whereby the eggs and larvae are transported to the nutrient rich southwest Cape waters. Though studies of ichthyoplankton in Natal waters are still in their infancy, the presence of large quantities of pilchard eggs has been confirmed (Anders, 1975). Likewise, many nursery grounds of the abovementioned winter migrants are known to be in Cape littoral waters, namely the elf and leervis (van der Elst, 1981) seventyfour (Ahrens, 1964) and the karanteen (Joubert, 1981).

A generalized overview thus indicates a feeding migration during which the nutrient input provided by pilchard shoals serves to build up the condition of pre-spawning predator stocks. Once in the warmer Natal waters, these species spawn, often in proximity to the Agulhas Current (van der Elst, 1976) hence promoting the dispersal of their progeny to richer nursery grounds. The adult fish themselves migrate back to Cape waters during spring, clearly shown for the elf by extensive tagging studies (van der Elst, 1983).

The sardine run is a complex phenomenon that requires further study. The first shoals are sighted off the Transkei coast during May each year, usually by commercial airline pilots. These shoals are composed primarily of the true Cape pilchard S. ocellata, though about 1% of the catches made by beach seine-net fishermen consists of the ubiquitous red-eye sardine Etrumeus teres. Later in the season the composition of 'sardine' shoals changes and catches made off Durban and northwards invariably comprise a mixture of filter-feeding fishes, some temperate, some tropical. Although the relative proportions of these may vary from shoal to shoal, the following species are usually present in association with the pilchard: red eye sardine (E. teres), maasbanker (Trachurus capensis), mackerel (Scomber japonicus), shortfin scad (Decapterus macrosoma) and the kelee shad (Hilsa kelee). Whereas the first three are of temperate origin, the latter three are tropical fishes, their presence in the 'sardine' shoals becoming progressively more prominent northwards and later in the season.

Sardines have been known to occur in Natal as late as December. This means that not only the winter migrants derive nutrients from these clupeid fishes, but they are known to be an important source of food for summer fishes also, most of which have made their appearance by that time (van der Elst and Collette, 1984). It is significant therefore, that the Natal region may serve as a 'contact' zone through which energy from high-productivity temperate regions is transferred to tropical regions. No doubt this input of energy contributes to the diversity and abundance of the Natal ichthyofauna as a whole.

Actual energy pathways within the Natal marine ecosystem are not known although aspects of intertidal energy budgets have been studied (Berry, 1982). It has been shown that decaying seaweed and terrestrial plant matter washed down rivers is macerated by wave action and provides a major input of nutrients to filter feeders such as mussels, oysters and ascidians (Schleyer, 1981). The high physical energy of the Natal surf zone promotes high biological productivity attained by these filter feeders, such that Berry (1978) recorded peak production of the brown rock mussel (*Perna perna*) exceeding 40 kg of flesh per square metre per annum. Moreover, the diversity and high biomass of carnivorous nearshore teleosts is clearly associated with this high mussel productivity (Berry *et al*. 1982). The abundance of early plantigrade stages and subsequent young mussel stocks coincides with the seasonal appearance of many other juvenile animals such as the pompano (*Trachinotus africanus*), the piggy (*Pomadasys olivaceum*), the catface rockcod (*Epinephelus andersoni*), the lampfish (*Dinoperca petersii*) (Berry *et al*. 1982) and also the rock lobster (*Panulirus homarus*) (Berry 1971). Many of these feed extensively on the small mussels during their juvenile phase before migrating to deeper waters and so transporting components of nearshore energy to shelf areas.

Clearly the Natal ichthyofauna belongs to an open ecosystem with pronounced seasonal changes, much of which can be directly related to physical variables such as temperature and current. This in turn influences migration and spawning patterns of the ichthyofauna, and four basic types can thus be identified:

a) resident fish that spawn locally and have local nurseries;

b) pelagic summer migrants that spawn in the tropics and have distant nurseries;

c) demersal summer migrants that spawn locally and have local nurseries;

d) winter migrants that are mostly endemic to South African waters, spawn locally and have Cape nurseries.

EXPLOITATION OF NATAL FISH

The Fishery

Obviously not all fish are of direct value to man, and the previous discussions on zoogeography, endemism and habitat preference become much more relevant when considered in relation to the actual exploited resource.

The FAO species manuals for the Western Indian Ocean (Area 51) lists 2 365 species of 144 families (including sharks but excluding batoids) as being of "interest to fisheries" (Fisher and Bianchi, 1984). Detailed descriptions are provided on species considered to be of particular commercial importance to man, and these number 955 species of 135 families. Analysis of their geographical distribution reveals that 501 (52%) of these commercially important species are encountered off Natal.

The distribution of commercial fisheries in the Indian Ocean was investigated by Rass (1965) who identified four main biogeographical fishery complexes, namely tropical coastal (off the coasts of East Africa and South Asia), tropical oceanic (open waters of the Indian Ocean), western south temperate (off the coast of South Africa) and eastern south temperate (off the coast of West Australia). Yet again the east coast of South Africa (including Natal) is singled out as a distinct region.

Despite the identity of these defined commercial regions and despite the numeric diversity of species, the Indian Ocean rates comparatively low in terms of total tons landed. No more than 5.5% of the world's marine catch comes from the Indian Ocean, though it occupies 20% of the world's ocean surface. The West Indian Ocean alone (Area 51) produced only 1.8 million metric tons or 3% of the world finfish catch in 1982 (FAO, 1982). If tunas and sharks are conservatively taken to be oceanic (which is not always the case) then 82% of the catch is based on species of the limited coastal or shelf waters, consisting primarily of the red fishes and rockcods (25%), the herrings, sardines and anchovies (24%) and the kingfishes, mullets and sauries (13%). Still less impressive is the total catch of Natal fish, which ranges from 1 500 to 2 000 m.t. of linefish and about 500 m.t. of herring-like fishes per annum (van der Elst and Garratt, 1984), a mere fraction of the total catch. Why then should Natal's fishes be credited with any importance?

The answer to this lies simply in the fact that total tonnage does not adequately describe the value of fish to man. The real asset of a resource should be seen in relation to the coastal communities exploiting it. Thus a 3 kg bream caught by a subsistence fisherman in his estuarine fishtrap is likely to be as valued as 300 kg of pilchards netted by a commercial trawlerman. Similarly, the value of fish as a recreational resource may outweigh the pure commercial or food value. This is certainly so in Natal where a considerable range of demands are placed on the fishery: economic, nutritional, social and recreational. All of these need to be reasonably satisfied although not all are compatible with each other and often require different management strategies.

The most valued of Natal's marine fish resources are the so-called linefish, which are those teleosts and elasmobranchs caught on hook and line and considered to be of value either as food or as a source of recreation. Though many are caught by the estimated 52 000 shore-based 'rock and surf' anglers, the greatest part of the linefish catch is made from craft at sea. These may be harbour-based lineboats, charter boats or strikers, some with permanent crew and freezing facilities (28 in total) or catches may be from the approximately 1 400 skiboats that launch directly from Natal's beaches. These skiboats are specially designed to cope with heavy surf conditions and this type of fishing has become very popular with recreational anglers, especially since about 1975.

Clearly there are many facets to linefishing in Natal and during 1983 the following number of fishermen were known to be involved.

recreational shore	52 000
recreational offshore	6 077
commercial offshore	510

To this should be added an estimated 30 000 people who were fare-paying occasional anglers on charter boats. This means that the total number of fishermen expecting a benefit from the Natal linefish resource probably lies between 75 000 to 100 000 per annum.

Trends in Natal's Fish Resources

The earliest serious fishing ventures in Natal took place around 1880, and the first angling clubs were established in 1885 (Robinson and Dunn, 1923). Initially commercial fishing was mostly in the form of beach seining for the elf (Pomatomus saltatrix) and the pilchard (Sardinops ocellata), both distinctly seasonal species. At the turn of the century steam-powered linefishing boats were introduced, reaching a maximum of 17 by 1932. Total landings amounted to about 1 100 tons per annum (Natal Fisheries Dept, 1907-1933). By 1983 the fleet had grown manyfold and its composition had changed significantly to include 5 line-boats, 23 charter boats and about 1 400 skiboats (van der Elst and Garratt, 1984).

Nevertheless, during this half-century the total landings have remained more or less constant. This has obviously meant a much smaller catch per unit effort for individual fishermen, amounting to an overall decrease of 85% for the catch per individual lineboat year (Fig. 9.1., van der Elst and Garratt, 1984). Of worse consequence was the drastic change in species composition of the catch (Fig. 9.1). Valued target species, such as the '74', steenbras, Scotsman and musselcracker, had all but disappeared, to be replaced by less-valued species, such as slinger, tuna and a large variety of 'other' species. Significantly, the overfished species are all endemic to South African waters, clearly a sign of their vulnerability. The slinger is presently the single most important species to the linefishery, but how long stocks will sustain the pressure remains to be seen, because it, too, is of limited distribution, being confined to southern Mozambique, south-east Madagascar and Natal (Bauchot and Bianchi, 1984; Garratt, 1984).

Like many other sparids the slinger undergoes sex reversal, being a protogynous process whereby female fish convert to males over a length range of 33 to 43 cm (Garratt, 1984). Because natural mortality limits the proportion of older fish in the population, it stands to reason that female slinger far outnumber males in the natural state - probably 3:1 (Garratt, 1984). Fishing mortality will aggravate this situation, and Garratt (1984) has demonstrated how this normal sex ratio can reach 19:1 in heavily exploited regions. Clearly, sex reversal could be an

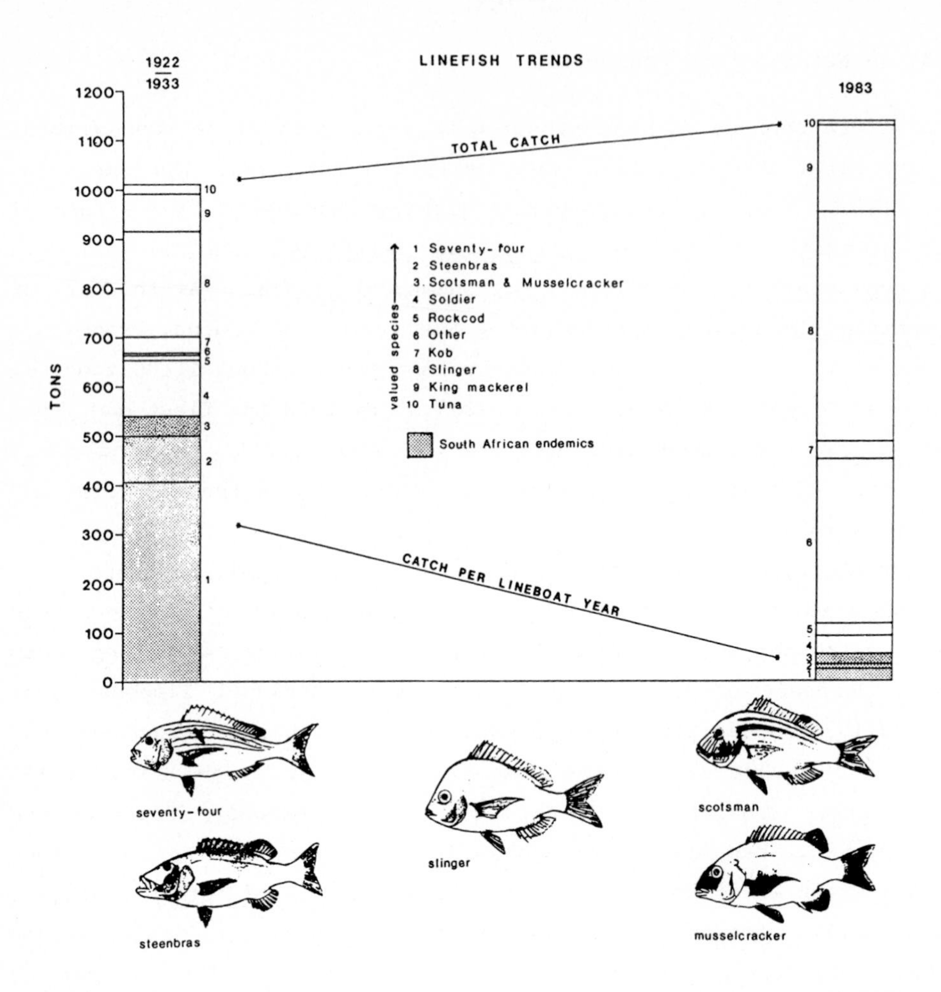

Figure 9.1 Changes in the catch composition, catch and CPUE in the Natal Fishery with illustrations of valued species. (Fish illustrations from: Fischer and Bianchi, 1984)

advantageous survival strategy because the greater the fishing pressure, the greater the proportion of females will become,so that population fecundity will be least impaired. If slinger were protandrous hermaphrodites (male to female), it would probably not have sustained the fishing pressure imposed on it this half-century. Indeed, the '74' does not undergo sex reversal (Ahrens, 1964), which is quite possibly a contributing factor to its marked decline in comparison with slinger.

Clearly the Natal linefishery is under stress. The fact that total landings have remained relatively constant probably allayed concern about the future of this resource and delayed the management actions that should have been taken much earlier. Assessment of stocks of multispecies tropical fisheries is never easy, but the association between changes in species composition and heavy harvesting is too obvious to be ignored (Larkin, 1982). What were taken to be natural fluctuations are now certain to be man-induced. The situation in Natal waters is particularly disconcerting because the change in composition has seen a reduction of the most valued fish. The fact that these are also principally endemic species, places a considerable responsibility on local fisheries management to reverse this negative trend. Whereas landings of the 1920s comprised 58% endemics by weight, present-day catches include only 9.5% of local species. Fish that now form a greater part of the catch are mostly wide-ranging and are usually also exploited by other countries in their own areas. It follows that local fisheries management will become less effective unless joint action is considered with neighbouring states.

Acknowledgements

Assistance and valuable comments were gratefully received from Professor A P Bowmaker, Dr A J de Freitas, P Garratt and F Adkin.

REFERENCES

AHRENS, R L (1964). A preliminary report on the biology of the seventy-four Polysteganus undulosus (Regan) in Natal waters. Unpublished M.Sc. thesis. University of Natal, Durban: 1-77.

ANDERS, A S (1975). Pilchard and anchovy spawning along the Cape east coast. **South African Shipping News and Fishing Industry Review,** 30(9), 53-57.

BAIRD, D (1971). Seasonal occurrence of the pilchard Sardinops ocellata on the east coast of South Africa. **Investigational Report. Division of Sea Fisheries (South Africa),** 96, 1-19.

BASS, A J, J D AUBREY and N KISTNASAMY (1973). Sharks of the east coast of southern Africa. I: The genus Carcharhinus (Carcharhinidae). **Investigational Report. Oceanographic**

Research Institute, 33, 1-168.

BERG, L S (1933). Die bipolare verbreitung der organismen und die Eiszeit. **Zoogeography** 1, 444-484.

BERRY, P F (1971). The biology of the spiny lobster Panulirus homarus (Linnaeus) off the east coast of South Africa. **Investigational Report. Oceanographic Research Institute,** 28, 1-75.

BERRY, P F (1978). Reproduction, growth and production in the mussel, Perna perna (Linnaeus) on the east coast of South Africa. **Investigational Report. Oceanographic Research Institute,** 48, 1-27.

BERRY, P F, R P VAN DER ELST, P HANEKOM, C S W JOUBERT, M J SMALE (1982). Density and biomass of the ichthyofauna of a Natal littoral reef. **Marine Ecology - Progress Series,** 10(1), 49-55.

BRIGGS, L S (1974). **Marine zoogeography.** New York, McGraw-Hill, 1-475.

CARCASSON, R H (1977). **A field guide to the coral reef fishes of the Indian and West Pacific Oceans.** London, Collins, 1-320.

CARTER, R and M SCHLEYER (1978). Zooplankton. In: HEYDORN, A E F ed. Ecology of the Agulhas Curent region. An assessment of the biological responses to environmental parameters in the southwest Indian Ocean, **Transactions of the Royal Society of South Africa,** 43(2), 167-170.

COHEN, D M (1973). Zoogeography of the fishes of the Indian Ocean. In: ZEITSCHEL, B ed. **The biology of the Indian Ocean.** Heidelberg, Springer-Verlag, 451-463.

EKMAN, S (1953). **Zoogeography of the sea.** London, Sidgwick and Jackson, 1-417.

FAO (1982). Yearbook of fishery statistics. Catches and landings, 54, 1-393.

FISHER, W and C BIANCHI (1984). FAO species identification sheets for fishery purposes, Western Indian Ocean (Fishing area 51). Rome, FAO, 6v.

GARRATT, P A (1984). The biology and fishery of Chrysoblephus puniceus (Gilchrist and Thompson, 1917) and Cheimerius nufar (Ehrenberg, 1830), two offshore sparids in Natal waters. Unpublished M.Sc. thesis, University of Natal, Durban, 1-139.

GARRATT, P A (1985). The offshore linefishery of Natal. I: Exploited population structures of the sparids Chrysoblepus puniceus and Cheimerius nufar. **Investigational Report. Oceanographic Research Institute,** 62, 1-18.

GARRATT, P A (1985). The offshore linefishery of Natal. II: Reproductive biology of the sparids Chrysoblepus puniceus and

Cheimerius nufar. **Investigational Report. Oceanographic Research Institute,** 63, 1-21.

GREENWOOD, P H, D E ROSEN, S H WEITZMAN and G S MYERS (1966). Phyletic studies of teleostean fishes with a provisional classification of living forms. **Bulletin of the American Museum of Natural History,** 131(4), 341-455.

HUBBS, C L (1952). Antitropical distribution of fishes and other organisms. **Seventh Pacific Science Congress,** 3, 1-6.

JOUBERT, C S W (1981). Aspects of the biology of five species of inshore reef fishes on the Natal coast, South Africa. **Investigational Report. Oceanographic Research Institute,** 51, 1-16.

LARKIN, (1982). Introduction. In PAULY, D and G I MURPHY. **Theory and management of tropical fisheries:** Proceedings of the ICLARM/CSIRO Workshop on the Theory and Managament of Tropical Multispecies Stocks, 12-21 January 1981, Cronulla, Australia. ICLARM Conference Proceedings, N°9, pp. 1-3.

MACARTHUR, R H and E O WILSON (1967). **The theory of island biogeography. Monograph in population biology. 1.** Princeton, Princeton University Press, 1-203.

McLACHLAN, G R and R LIVERSIDGE (1978). **Roberts birds of South Africa;** 4th ed. Cape Town, John Voelcker Bird Book Fund, 1-659.

MYERS, G S (1941). The fish fauna of the Pacific Ocean, with especial reference to zoogeographical regions and distribution as they effect the international aspects of fisheries. **Proceedings of the Pacific Science Congress,** 3, 201-210.

NELSON, J S (1976). **Fishes of the world.** New York, John Wiley, 1-416.

ODA, D K and J D PARRISH (1981). Ecology of commercial snappers and groupers introduced to Hawaiian reefs. In: GOMEZ, E D, E D BIRKELAND, C E BUDDEMEIER, R W JOHANNES, J A MARSH and R T TSUDA eds. The reef and man. Proceedings of the fourth international coral reef symposium: v.1. Quezon City, Marine Science Center, University of Philippines, 1-735.

RASS, T S (1965). Commercial ichthyofauna and fisheries resources of the Indian Ocean. **Trudy Instituta Okeanologii. Akademiya Nauk, USSR,** 80, 3-31.

ROBINSON, R and J S DUNN (1923). **Salt water angling in South Africa.** Durban, Robinson, 1-315.

ROSS, G J B (1984). The smaller cetaceans of the east coast of southern Africa. **Annals of the Cape Provincial Museum,** 15(2), 173-410.

SCHLEYER, M H (1981). Microorganisms and detritus in the water column of a subtidal reef of Natal. **Marine Ecology - Progress Series,** 4,

307-320.

SMITH, J L B (1965). **The sea fishes of South Africa:** 5th ed. Cape Town, Central News Agency, 1-580.

SMITH, M M (1980). Marine fishes of Maputaland. In: BRUTON, M N and K K H COOPER, eds. **Studies on the ecology of Maputaland.** Grahamstown, Rhodes University, 164-187.

STEINITZ, H (1973). Fish ecology of the Red Sea and Persian Gulf. In: ZEITZSCHEL, B ed. **The biology of the Indian Ocean.** Heidelberg, Springer-Verlag, 465-466.

VAN DER ELST, R P (1976). Game fish of the east coast of southern Africa. 1. The biology of the elf, Pomatomus saltatrix (Linnaeus) in the coastal waters of Natal. **Investigational Report. Oceanographic Research Institute,** 44, 1-59.

VAN DER ELST, R P (1980). The marine sport fishery of Maputaland. In: BRUTON, N M and K H COOPER eds. **Studies on the ecology of Maputaland.** Grahamstown, Rhodes University, 188-197.

VAN DER ELST, R P (1981). **A guide to the common sea fishes of Southern Africa.** Cape Town, Struik, 1-367.

VAN DER ELST, R P (1983). Studies of the elf (Pomatomus saltatrix) - Is there a lesson for the Marine Linefish Programme? 5th National Oceanographic Symposium, Grahamstown, 24-28 Jan. 1983. **South African Journal of Science,** 79(4): 166 Abstract.

VAN DER ELST, R P, R J M CRAWFORD and P A SHELTON (1978). Marine teleost fishes. In: HEYDORN, A E F ed. Ecology of the Agulhas Current region. An assessment of biological responses to environmental parameters in the southwest Indian Ocean. **Transactions of the Royal Society of South Africa,** 43(2), 179-183.

VAN DER ELST, R P and B B COLLETTE (1984). Game fishes of the east coast of southern Africa. 2. Biology and systematics of the queen mackerel Scomberomorus plurilineatus. **Investigational Report. Oceanographic Research Institute,** 51, 1-12.

VAN DER ELST, R P and P A GARRATT (1984). Draft management proposals for the Natal deep reef linefishery. Unpublished report: **Oceanographic Research Institute,** 1-30.

WALLACE, J H (1975a). The estuarine fishes of the east coast of South Africa. I. Species composition and length distribution in the estuarine and marine environments. II. Seasonal abundance and migration. **Investigational Report. Oceanographic Research Institute,** 40, 1-72.

WALLACE, J H (1975b). The estuarine fishes of the east coast of South Africa. III. Reproduction. **Investigational Report. Oceanographic Research Institute,** 41, 1-51.

WALLACE, J H and R P VAN DER ELST (1975). The estuarine fishes of the east coast of South Africa. IV. Occurrence of juveniles in estuaries. V. Ecology, estuarine dependence and status. **Investigational Report. Oceanographic Research Institute,** 42, 1-63.

WALLACE, J H, H M KOK, L E BECKLEY, B BENNETT, S J M BLABER and A K WHITFIELD (1984). South African estuaries and their importance to fishes. **South African Journal of Science,** 80(5), 203-207.

WYRTKI, K (1971). Oceanographic atlas of the international Indian Ocean expedition. National Science Foundation. Washington, D.C., **US Government Printing Office,** 1-531.

Chapter 10

POLLUTION AND EFFLUENT DISPOSAL OFF NATAL

Allan D Connell
National Institute for Water Research
Council for Scientific and Industrial Research

INTRODUCTION

The marine environment of Natal has a relatively short history of domestic and industrial pollution. Sugar mills have been responsible for the pollution of a number of estuaries, but these effects have been localized and have not influenced the marine environment directly. Nevertheless, the interdependence of marine and estuarine systems is recognized.

Sewage disposal to sea at Durban began as early as July 1896, after an announcement in the *Natal Mercury* on March 3, 1891, outlined the plan. Subsequent editions of the newspaper carried numerous readers' letters criticizing sea disposal (Lynsky, 1982)! Sewage was screened and then discharged from the North pier through the outfall into the harbour channel (Figure 10.1), during the first few hours of the ebb tide. This arrangement, and later, larger versions, remained the principal means for waterborne sewage disposal for many years, and by 1968 the volume discharged had grown to 90 000 m^3/day. In addition, because of the growing population on the Bluff, 20 000 m^3/day was discharged into the surf at Fynnlands. In 1968 and 1969 this poor state of affairs was corrected when two outfalls were constructed to discharge sewage via submarine pipelines.

A bacterial survey on the distribution and occurrence of coliforms and pathogenic indicators of sewage pollution in the vicinity of Durban prior to the commissioning of these pipelines indicated that 15 out of 28 stations located between the Umgeni river and Isipingo were heavily polluted (Livingstone, 1969). A resurvey of the same 28 stations after the pipelines became operational showed, from 10 sampling runs, an average of four stations to be heavily polluted (Livingstone, 1976).

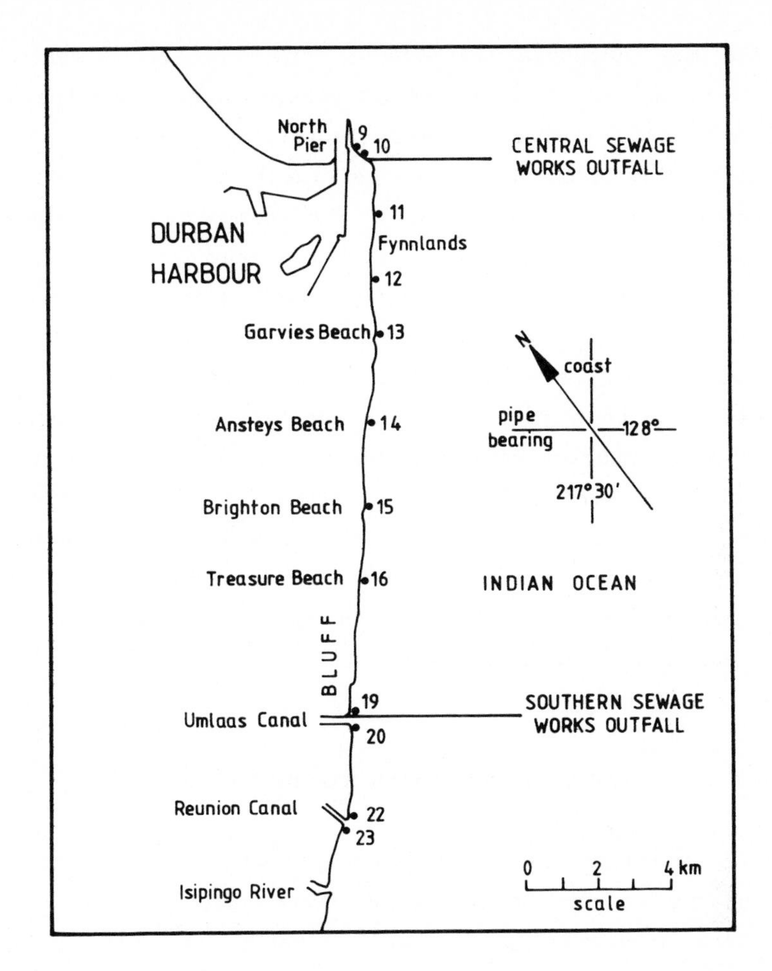

Figure 10.1 Sampling points along the Durban "Bluff" beaches, where mussels were collected for polycyclic aromatic hydrocarbon (PAH) analysis (Butler and Sibbald, 1986).

The results of these and other surveys have led to the development of a sophisticated and stringent system of microbial classification of seawater bathing beaches, which has been tested against the experience and knowledge of parochial epidemiology of the local medical officer of health (Mackenzie and Livingstone, 1983).

The major marine discharges of industrial effluents in Natal have a much shorter history. Amongst the oldest industrial discharges is that from the AECI factory at Umbogintwini, south of Durban (see Figure 10.2). This discharge has, however, never been of great volume and even today is only some 3 000 to 4 000 m^3/day. In 1959 plans were underway for two industrial pipelines, one to carry titanium oxide wastes via a 1.5 km pipeline to sea off Umbogintwini, and the other to deal with the effluent from a rayon factory at Umkomaas (see Figure 10.2), which from 1955 discharged effluent into the surf on the southern bank of the Umkomaas river. The former began discharging in early 1962, at a rate of about 1 800 m^3/day and the latter was commissioned in March 1967, discharging at a rate of 80 000 m^3/day.

Since completion of the two Durban pipelines, no other major pipelines were laid off the Natal coast until late 1983, when construction began on a double pipeline at Richards Bay (Figure 11.1), one line to carry buoyant effluent and the other to discharge a dense effluent-containing gypsum. The two pipes, lying virtually side by side, were pulled out to sea in early 1984 and effluent was first discharged through the buoyant effluent line in October 1984. Problems with pumps and pipes delayed the discharge of gypsum from the dense line until June 1985 (see also Chapter 11).

SPECIFIC POLLUTANT LEVELS IN THE NATAL MARINE ENVIRONMENT

In 1974 a research programme was initiated to investigate the pollution status of the Natal marine environment (Cloete and Oliff, 1976). This programme concentrated largely on trace metals and persistent organochlorine compounds.

Trace Metal Levels

No published information on trace metals is available other than that reported by Cloete and Oliff (1976) and Gardner, et al. (1983). Water and sediment samples collected on coastal reference transects (Figure 10.2), each consisting of three stations sited on the 25, 50 and 100 m depth contours, have yielded considerable data which are presently being processed. When compared with data collected further offshore from a set of stations off Durban extending to 100 km offshore, these data show that little coastal influence is discernible except in the case of cobalt and nickel and, to a lesser extent, lead (Table 10.1).

Regular surveys have been made of metal levels in samples of mussels,

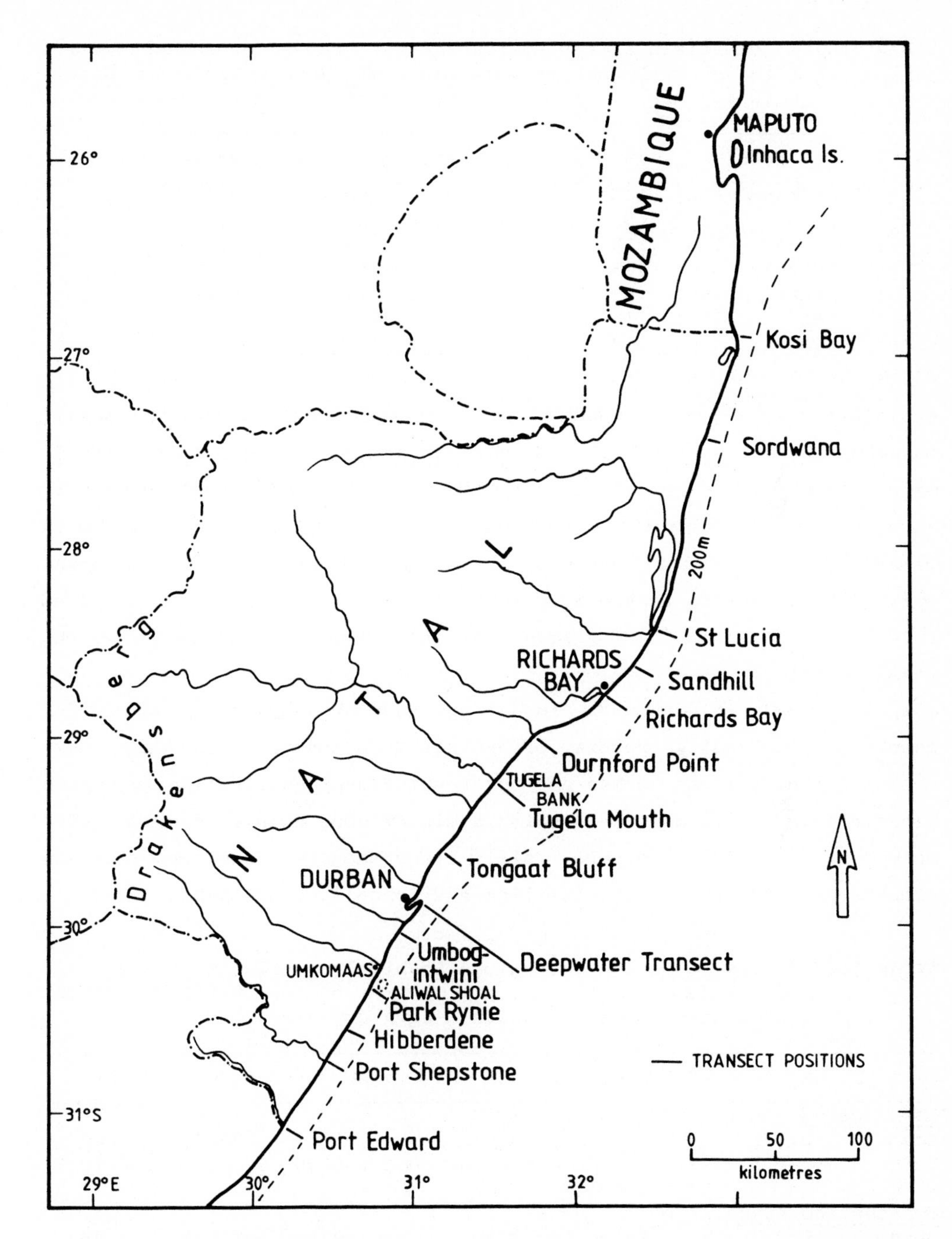

Figure 10.2 The Natal coast, showing thirteen coastal reference transects and the deepwater (ECOR) transect off Durban.

Table 10.1 Comparison of median trace metal values for inshore waters (100 m depth contour) and offshore waters off Natal. (Units are in µg/l and n denotes the number of samples).

Metal		n	Inshore Waters	n	Offshore Waters
Cadmium	(Cd)	114	0.069	100	0.070
Cobalt	(Co)	61	3.80	48	0.924
Copper	(Cu)	119	0.650	124	0.560
Mercury	(Hg)	103	0.048	113	0.056
Nickel	(Ni)	40	1.17	20	0.054
Lead	(Pb)	116	0.59	114	0.373
Zinc	(Zn)	118	1.94	116	1.90

oysters and rock lobsters collected at various points along the coast. Mussels (*Perna perna*) and oysters (*Crassostrea cuccullata* and *C. margaritacea*) have proved most suitable (Table 10.2), and generally show low levels of metals. An exception was the level of chromium in mussels from the Bluff beaches of Durban (Figure 10.1), in which, during 1983, levels ranged from 40.8 to 167 µg/g dry mass with a mean (n = 9) of 79.0 µg/g. The source of the chromium appeared to be the Umlaas Canal.

Metal levels in sediments in Durban Harbour have given cause for concern in the past. Transfer of oysters from Knysna on the Cape South coast (Figure 1.1) to Durban and their subsequent cultivation in Durban Harbour over twelve months revealed a significant drop in cadmium but a marked rise in zinc, copper, lead, iron, manganese and chrome levels within the first two months (Watling and Connell, unpublished data).

Table 10.2 Toxic metal concentrations in bivalve molluscs from various sites between Tongaat and Sezela on the Natal coast but excluding Durban. (Units = µg/g dry mass).

	Hg	Cu	Cd	Pb	Zn	Cr
Perna perna						
Mean	0.071	11.8	0.89	0.80	93.5	23.8
Range	0.042-0.112	2.49-78.0	0.05-1.85	0.005-5.39	50.2-283	1.6-90.5
Crassostrea cuccullata						
Mean	0.217	179	2.29	0.997	1 000	6.74
Range	0.055-0.291	32.0-438	0.60-4.77	0.005-5.30	-	1.6-19.7
Crassostrea margaritacea						
Mean	0.142	191	4.89	3.75	1 000	18.2
Range	0.051-0.329	42.9-546	1.96-10.0	0.005-9.00	-	0.2-58.4

Table 10.3 Chlorinated hydrocarbon levels in mullet (Mugil sp.p) from the estuaries of Natal. Sites given can be found on Figures 1.1 and 10.3. (All data in µg/kg wet mass).

Location	DDT	DDE	Dieldrin	PCB's
Kosi Bay (1976)	9 - 860	4 - 270	1 - 44	ND
Kosi Bay (1981)	ND	ND	ND	ND
St Lucia Estuary	1 - 38	0.5 - 79	0.5 - 8	ND
Richards Bay (1976)	ND - 170	1 - 80	ND - 4	ND
Richards Bay (1981)	ND	ND	ND	ND
Mgeni Estuary (Durban North)	8 - 205	ND - 141	6 - 169	ND
Durban Bay (1978)	9 - 36	ND - 31	ND	ND
Durban Bay (1978*)	24 - 119	15 - 279	ND - 595	ND
Umgababa	2	2 - 3	ND	ND
Transkei				
Port St Johns	ND - 30	ND - 20	ND	ND
Mbashe Estuary	ND - 17	ND - 34	ND	ND

*An opportunistic sample of large mullet found trapped in the dry dock

Pesticides

Three chlorinated hydrocarbons are regularly encountered in various nearshore faunal species along the coast of Natal. These are DDT, (and its metabolites TDE and DDE), dieldrin and the polychlorinated biphenyls (PCBs).

The initial surveys, conducted between 1978 and 1980, indicated that DDT derivatives and dieldrin were readily detectable in fauna of estuaries in the vicinity of Durban, East London and Port Elizabeth, with the highest levels in the vicinity of Durban. For mullet, Mugil sp. these levels generally tended to range from 50 to 200 µg/kg wet mass (Table 10.3). Away from these centres, levels tended to be lower, ranging from below detection limits to about 30 µg/kg. It was thus shown that contamination was localized to usage areas. In South Africa use of the two pesticides DDT and dieldrin was banned in 1976 (Darracott, 1975), although DDT is still being used in malarial areas of KwaZulu, and particularly in Maputaland. However, provided this is used carefully on house walls only, it should not be a major cause of environmental contamination.

Dolphins have been examined for chlorinated hydrocarbon content, as they are at the pinnacle of the marine food chain and their blubber is

Table 10.4 Chlorinated hydrocarbon levels in dolphins from Natal (L = liver, B = blubber, ND = not detectable). (All data in µg/kg wet mass).

Locality	Date	DDT	DDE	TDE	Dieldrin	PCBs
Tursiops aduncus (blubber analysis only).						
Natal	11/1977	2 700	8 300	3 000	350	14 370
Natal	11/1977	750	6 190	350	31	15 950
Natal	11/1977	930	1 230	590	150	2 580
Natal	11/1977	1 370	3 160	1 250	390	1 760
Natal	11/1977	4 870	7 117	1 540	520	30 567
Scottburgh	7/1978	2 200	24 100	ND	ND	21 950
Scottburgh	7/1978	1 730	12 900	ND	ND	15 260
Scottburgh	7/1978	1 286	28 100	ND	ND	19 750
Salt Rock	7/1978	2 050	7 670	ND	ND	21 850
Natal Coast	12/1979	6 880	40 100	ND	856	47 000
Natal Coast (calf)	12/1979	189	623	ND	8	ND
Natal Coast	12/1979	6 050	18 300	ND	329	42 800
Natal Coast	12/1979	5 690	23 400	ND	1 210	26 100
Durban	12/1979	4 950	13 600	ND	1 900	22 900
Natal Coast	1981	ND	448 000	ND	ND	46 900
Natal Coast	1981	ND	14 100	ND	597	ND
Natal Coast	1981	ND	1 920	ND	346	9 910
Natal Coast	1981	1 900	1 690	ND	431	ND
Natal Coast	1981	5 580	13 400	ND	ND	ND
Natal Coast	1981	613	1 000	ND	361	ND
Natal Coast	1981	2 830	6 370	ND	ND	11 500
Natal Coast	1981	10 200	23 400	ND	611	61 900
Natal Coast	1981	6 930	15 100	ND	2 410	50 100
Sousa plumbea (blubber and liver analyses)						
Natal (L)	12/1979	515	759	59	194	4 290
Natal (B)	12/1979	16 700	31 500	428	235	130 000
Richards Bay (L)	1981	ND	55	ND	ND	ND
Richards Bay (B)	1981	6 970	10 900	ND	310	46 700
Richards Bay (L)	1981	ND	72	ND	26	ND
Richards Bay (B)	1981	ND	2 020	243	248	6 600
Richards Bay (L)	1981	ND	197	ND	18	1 000
Richards Bay (B)	1981	1 890	3 020	ND	1 130	13 800
Richards Bay (L)	1981	ND	23	ND	ND	ND
Richards Bay (B)	1981	410	452	ND	14	ND

a major absorber of these compounds. Levels have been found to be very variable (Table 10.4), but with a general trend of higher levels in the dolphins of Natal than in those from the eastern Cape (A C Butler, pers. comm.).

Petroleum Hydrocarbons

Until recently no information has been available on oil levels in the seas off Natal, apart from tar-ball data (Shannon et al. 1983).

Two studies were recently made, one on the oils of the surface microlayer off Richards Bay, and the other on oils in mussels in the vicinity of Durban. In the former study (Butler and Sibbald, 1985) a Teflon (PTFE) disc was applied momentarily to the ocean surface, and the hydrocarbons adhering to the surface were then collected for gas-chromatographic analysis in the laboratory.

The technique proved to be very simple and efficient. Levels of petroleum hydrocarbons were generally at least an order of magnitude higher (mass per unit area) than those reported in the literature for studies in which wire-mesh screens were used for sampling. This was to be expected since wire-mesh screen sampling results in the analysis of a water film of finite depth, while the present method scavenges only the compounds at the air-water interface (A C Butler, pers. comm.). Of the six samples collected, five contained petroleum-based compounds, and two appeared to comprise the same oil. All petroleum found appeared to be fuel oil, the presence of measurable levels of n-alkanes showed that the oil had not been present in the marine environment for a sufficient length of time for any marked biodegradation to have taken place. Levels detected are given in Table 10.5.

The absence of crude oil fractions was an encouraging indication of the minor importance of this source of pollution to the Natal coast at the time of the study (October 1983).

The mussel study was done in conjunction with the disposal of sludge to sea experiment at Durban, and was the first use of the mussel watch concept for coastal oil pollution in South Africa (A C Butler, pers. comm). Sampling was confined to the stretch of coastline from Durban

Table 10.5 Levels of n-Alkanes. Units = $mg.m^{-2}$. UCM = unresolved complex mixture PAH = polycyclic aromatic hydrocarbons.

Sample	2	3	4	5	6
Σn-Alkanes	3.92	2.38	12.40	1.11	5.25
Σ(n-Alkanes + Alkane UCM)	24.10	1.22	79.20	4.13	25.10
PAH UCM	1.56	ND	4.94	0.25	3.32

mouth to Isipingo in the south, a distance of about 10 km (Figure 10.1). Evidence of contamination was obtained from mussels from all stations, with peak concentrations at stations 19 and 22, the latter being at the outlet of a drainage system serving the site of a large oil refinery. Knutsen and Sortland (1982) quote modified figures of various studies in Norway for total PAHs which were modified to wet-mass basis for comparison with the present results.
Their broad divisions were

Clean areas	75 - 375 ppb (µg/kg)
Urban areas	75 - 1 500 ppb
Industrial areas	375 - 3 750 ppb

Knutsen and Sortland's results for total resolved PAHs in mussels from "moderately polluted" parts of the Norwegian coast ranged from 75 to 1 900 ppb. In the present study total resolved PAHs were present at concentrations of 170 to 3 700 ppb, which when adjusted for the unresolved complex mixture (UCM), gave total PAHs of 350 to 19 500 ppb.

Curiously, the levels of UCMs at stations 19 and 20 were very similar, but were quite different from those of stations 22 and 23. The levels of UCMs at these two latter stations were again very similar, suggesting two different sources for the petroleum in mussels from stations 19-20 and 22-23, at the mouths of the Umlaas and Reunion canals, respectively.

DUMPING AT SEA

The only materials routinely dumped off the Natal coast are dredge spoils derived from the two harbours, namely, Durban and Richards Bay.

Maintenance dredging in Durban Harbour removes annually about 175 x 10^3 m^3 from within the harbour. This is dumped at a site 4 to 6 km offshore on a bearing 110 °T from the southern breakwater, where water depth exceeds 50 m. This material is fairly high in content of metals such as copper, zinc, lead, mercury and chrome (Gardner *et al*. 1983). A comparison with levels of metals in sediments from similarly fine muddy estuary substrates (Table 10.6) suggests that the metals do not necessarily all originate from harbour activities. The highest levels

Table 10.6 Metals in estuary and harbour fine sediments from along the Natal coast. All data in µg/g dry mass except for mercury, which is expressed as ng/g dry mass. (Data from Gardner *et al.* 1983 and Connell *et al.* 1985).

Locality	Date	n	Hg	Cu	Fe	Zn	Pb	Cd	Cr
Umgababa Estuary	1976	10	55	2.5	5 659	64	4.5	0.32	11
Durban Bay	1978	7	299	34	19 669	117	58.4	0.95	53
Richards Bay	1974	5	22	9.9	23 640	98	24.0	0.57	22
Richards Bay	1976	6	20	26.5	6 738	102	17.8	0.99	86
Richards Bay	1981	8	181	20.5	-	55	9.7	0.65	69
St Lucia	1978	8	21	35.6	37 300	47	9.4	0.23	102

of both copper and chromium (Table 10.6) were in the unindustrialized St Lucia estuary.

An additional 650 x 10^3 m^3 of sand are removed annually from the south-east of the south pier and dumped in an area about 6 km offshore bearing 135°T from the south pier. This is, however, clean sand which builds up in the inshore area as a result of the northward movement of sand along the Natal coast (Swart, 1983). In 1983 about 600 x 10^3 m^3 of sand were pumped onto Durban beaches in a period of six weeks, and presently about 200 x 10^3 m^3 per annum are being used to replenish the beaches of Durban.

Richards Bay Harbour was built between 1972 and 1976, and opened to shipping in April 1976. Massive amounts of fine sediment were pumped out of the harbour area onto beaches; the volume was estimated to exceed 50 x 10^6 m^3 (McClurg *et al.* 1985). In addition, dredge spoil was dumped at sea, and this dumping is continuing at a rate of between one and two million cubic metres per annum. Much of this consists of sand taken from south of the southern breakwater, where deposition of northward-moving sand has greatly widened the beach. Presently about 850 x 10^3 m^3 are pumped annually onto the recreation beach to the north of the harbour entrance, as this has become seriously eroded. The fine sediments removed from inside the harbour and dumped offshore, amount to about 100 x 10^3 m^3 per annum. Results of the analyses of some of these sediments are also presented in Table 10.6. The dump site was moved in May 1984 to a position some 4 km offshore in water about 25 m deep (McClurg *et al.* 1985).

POINT SOURCES OF EFFLUENT TO THE MARINE ENVIRONMENT OF NATAL

Effluent is being discharged to sea at 47 recognized points on the Natal coast. At 22 of these the amount discharged exceeds 10 000 m^3/day and at two it exceeds 100 000 m^3/day; the total volume along the whole coast exceeds 600 000 m^3/day. The data are presented schematically in Figure 10.3, and some notes on the discharges are presented in the Appendix.

THE RELATIVE POTENTIAL ENVIRONMENTAL IMPACT OF THE MAJOR PIPELINES AND OTHER EFFLUENT SOURCES IN NATAL

In order to enable the toxicities of the various discharges into the sea off Natal to be compared, toxicity testing techniques have been developed using local species as test organisms. The species used include the estuarine amphipod *Grandidierella* (Connell and Airey, 1982) and, more recently, eggs and larvae of the tropical marine pomacentrid fish *Dascyllus trimaculatus*. The newly fertilized embryos of the brown mussel *Perna perna*, and a sperm cell test with sea urchin sperm as described by Dinnel *et al.* (1982) have also been used.

Numerous tests on effluents have recently been conducted using the eggs of *Dascyllus*, to compare the toxicities of the major sea discharges along the Natal coast. The results of some of these toxicity tests are presented in Table 10.7. Those pipelines for which the difference between the theoretical worst-case dilution and the no effect toxicity test dilution are the greatest, are suggested as being those most likely to cause environmental damage. Of these the Felixton paper mill discharging presently into the surf has destroyed intertidal fauna and flora for several hundred metres on the adjacent beach. Currently, there is debate concerning the best method of dealing with this problem, particularly in view of plans to expand production at the plant.

In the case of S A Tioxide which shows the biggest contrast (see Table 10.7), diving surveys have shown only limited damage to benthic communities within 25 m of the pipe-end. Installation of an efficiently designed diffuser system will undoubtedly improve the

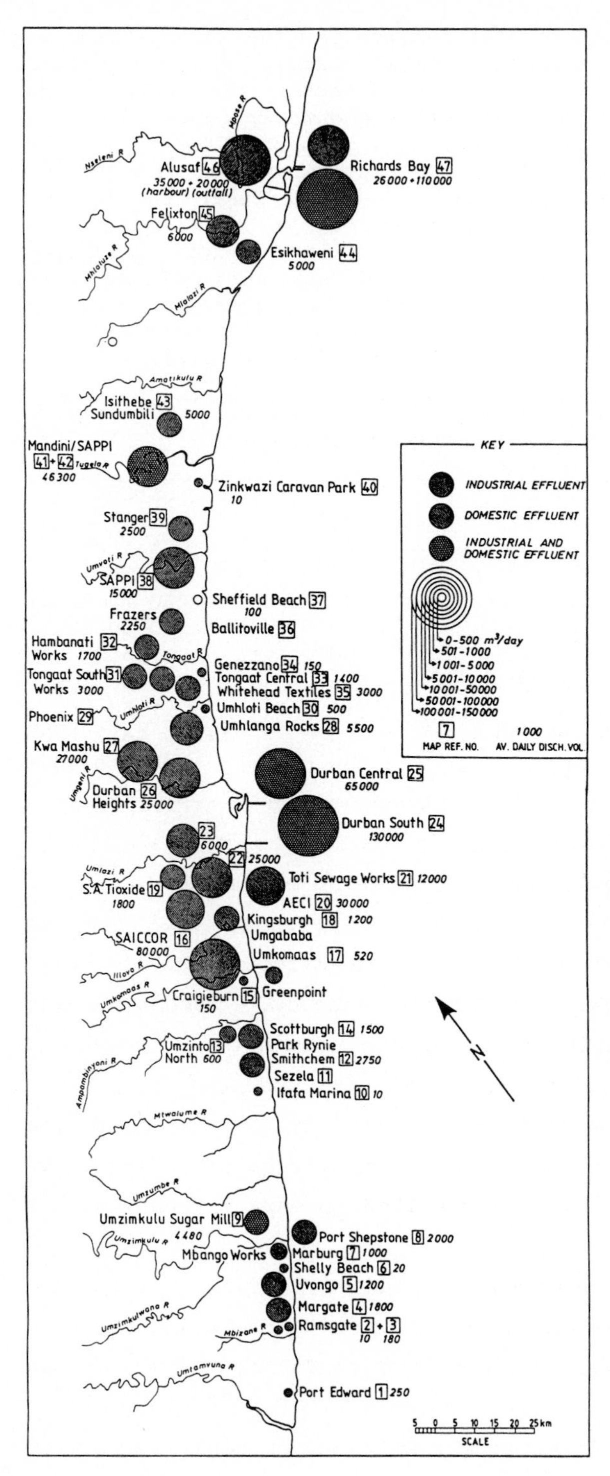

Figure 10.3 Schedule of marine and estuarine effluent disposal along the Natal coast, January 1985. (Updated, after Lang 1982). The map reference number (MR) is blocked, with further details given in the Appendix.

Table 10.7 Dilution and toxicity test results for the major effluents discharged to sea along the Natal coast. The theoretical worst-case dilution is calculated taking into account pipe design, effluent characteristics and water depth under "no current" conditions: this is the mean dilution across the boil (G Toms, pers. comm.). The two values for SAICCOR are without and with the present diffuser section in operation. Toxicity is determined on the basis of no effect dilution to eggs and larvae of _Dascyllus trimaculatus_ i.e. the lowest dilution of effluent in seawater at which hatching and development of larvae compares favourably with control volumes of pure seawater; Durban central was found to be somewhat variable.

Effluent	Seawater discharge depth (m)	Volume m^3/day	Theoretical worst case dilution	Toxicity
Smithchem	2-4	2 000	100	1:250
SAICCOR	25	80 000	20 (120)	1:250
S A Tioxide	30	1 700	20	1:1 250
AECI	2-4	3 000	100	1:50
Durban South	54-64	130 000	340	1:250 to 500
Durban Central	48-53	65 000	300	1:100 to 500
David Whitehead	2-4	1 500	100	1:500
SAPPI Stanger	-	15 000	N/A	1:50 to 100
SAPPI Mandini	-	45 000	N/A	1:50
Mondi Felixton	2-4	5 000	50	1:500 to 1 500
Richards Bay "A"	29	110 000	250	1:250 to 500

situation (G Toms, pers. comm.), and this is being recommended.

Of the major volume pipelines discharging at present, only that of SAICCOR gives a dilution less than the no-effect toxicity test dilution (Table 10.7), and thus warrants close study. Recent surveys have not revealed any ill effects to benthic communities around the pipe-end, although it should be noted that the effluent has a low specific gravity and is very warm (45 to 55°C), and thus rises to the surface very rapidly. Surveys of meiofauna on beaches most frequently contaminated by this effluent have also failed to show any detrimental effects.

Other studies presently in progress or planned include photographic surveys of reefs to record growth of corals, soft corals and sponges, and settlement studies near the pipe end to study the impact of the effluent in the pelagic zone.

It is suggested that if these potentially worst cases cannot be shown to be environmentally damaging, it can be assumed that all the major pipelines on the Natal coast are functioning satisfactorily. This has been shown to be the case with the Durban Central and Southern pipelines following a recent intensive study (Livingstone, 1983).

COMPARISON OF NATAL WITH OTHER PARTS OF THE WORLD

Many coastal regions of the world have suffered damage because of discharge of wastes into the ocean from urban and industrial complexes. Generally the problem has been localized and has resulted from the tendency for coastal development to occur in protected areas. Poor exchange rates in such areas lead to the assimilative capacity being rapidly exceeded as loads increase (Segar et al. 1984).

The Natal coast is remarkably straight and devoid of embayments; in addition there are no offshore islands. The continental shelf is also relatively steep (Chapter 2) and swept by the powerful Agulhas current. Effluent mixing rates and dispersion are therefore generally good, an example being the sludge disposal from two pipelines off Durban (Livingstone, 1983). Monitoring of sediments around the ends of both pipelines showed no signs of sludge deposition or benthic fauna enrichment. This is not surprising since in the area of the diffusers of these two pipelines, there is marked sediment movement, resulting in undermining of the pipeline which has had to be countered by draping netting over the line in order to trap sand (D C MacLeod, City Engineer Durban, pers. comm.).

In terms of effluent oxygen demand, the Natal coast is comparatively little affected, with one exception, namely that imposed by the cellulose mill at Umkomaas (SAICCOR). This mill produces more than twice the chemical oxygen demand (COD) that is produced by all the other pipelines and major discharges on the Natal coast (Table 10.8).

Anderson et al. (1979) mention that the worst case of pulp and paper mill impact during the early 1970s was in the United States in the Port Gardner area of Puget Sound, where, at maximum discharge, four mills produced a total of nearly 540 tons of five-day biological oxygen demand (BOD_5) per day.

This led to the ECOBAM study, started in 1972 and designed to

Table 10.8 Comparison of oxygen demand inputs of Natal pipelines, with some other areas of the world. The SAICCOR BOD value is based on a February, 1986, BOD of 4 715 mg/l (mean of 2x24 hour composites). The New York Bight values are from Anderson *et al.* (1979) and the Californian values from Segar *et al.* (1984). Data in metric tons/day.

	BOD	COD	SS
SAICCOR	377	1 640	29
AECI	-	0,9	-
Durban Southern Works	-	326	84
Durban Central Works	-	84,5	21
SAPPI Stanger	-	35-70	36
SAPPI Mandini	-	27	4
Mondi Felixton	-	50-75	5
Mondi Richards Bay	-	126	8
Triomf Richards Bay	-	-	5 200
New York Bight	1 500	10 000	
Southern California Bight	685	2 140	

measure the environmental impact of this effluent. By 1979 BOD_5 load had been reduced to approximately 136 metric tons per day. Although SAICCOR alone produces 377 tons of BOD_5 per day, the only significant impact thus far has been related to aesthetics including both colour and a tendency for the effluent to stabilize foam. This is further evidence of the offshore dynamics of this part of the Natal coast. However, the quality of this effluent is uncommonly poor. Anderson *et al.* (1979) list kraft mills as having a BOD_5 of 20 to 300 mg/l and a COD of 100 to 1 200 mg/l. Currently, SAICCOR effluent BOD_5 exceeds 4 000 mg/l, and that of COD is about 20 000 mg/l, whereas the new pulp and papermill at Richards Bay has been limited in its discharge permit, to a COD of 2 000 mg/l. The SAICCOR factory was built in the mid-1950s and is currently being partially modified so that the quality of the effluent will be improved (see Appendix 1, MR 16). It is hoped that the means will be found to update the entire mill as soon as possible, to bring it in line with modern technology and effluent discharge standards.

Acknowledgements

The assistance of the staff of the Department of Water Affairs in

Durban, in updating the Inventory of Point Source Outfalls on the Natal coast, which was originally compiled by Mr John Lang for the Natal Town and Regional Planning Commission, is gratefully acknowledged.

This contribution is published with the permission of the Director of the National Institute for Water Research.

REFERENCES

ANDERSON A R, P W ANDERSON, W M DURSTON, L L FALK, J E KERRIGAN, C D ROSE and R TOLLEFSON (1979). Sources. In: **Assimilative capacity of US Coastal Waters for pollutants.** Crystal Mountain. Proceedings of a Workshop.

ANONYMOUS (1984). Requirements for the purification of waste water or effluent. **South African Government Gazette No. 9225/991,** 18 May 1984.

BUTLER, A C and R R SIBBALD (1985). Petroleum hydrocarbons in the ocean surface microlayer off Richards Bay. In: **Environmental studies at Richards Bay prior to the discharge of submarine outfalls: 1974-1984.** (Eds: A D CONNELL, T P McCLURG and D J LIVINGSTONE), National Institute for Water Research, Durban. pp. 255-276.

BUTLER, A C and R R SIBBALD (1986). Aliphatic and polycyclic aromatic hydrocarbons in brown mussels, Perna perna from the Bluff, Durban, South Africa. **Science of the Total Environment** (in press).

CLOETE, C E and W D OLIFF (1976). South African marine pollution survey report 1974-1975. **South African National Scientific Programmes Report No. 8.** 60 pp.

CONNELL, A D (1983). Effluent Toxicity. In: **Detailed Report. Sludge disposal to sea.** Report No. 9, 137-143. City Engineer's Department, City of Durban, September 1983.

CONNELL, A D and D D AIREY (1982). The chronic effects of fluoride on the estuarine amphipods Grandidierella lutosa and G. lignorum. **Water Research** 16, 1313-1317.

CONNELL A D, T P McCLURG and D J LIVINGSTONE (Eds.) (1985). **Environmental studies at Richards Bay prior to the discharge of submarine outfalls: 1974-1984.** 290 p. Durban, September 1985.

DARRACOTT, A (1975). Pesticides in the marine environment 3. Pollution sources, research legislation and other control methods in South Africa. **Environment RSA** 2, 3-5.

DINNEL P A, Q J STOBER, S C CRUMLEY and R E NAKATANI (1982). Development of a sperm cell toxicity test for marine water. Aquatic Toxicology and Hazard Assessment : Fifth Conference. ASTM STP 766 J G Pearson, R B Forster and W E Bishopp Eds. pp. 82-98.

GARDNER B D, A D CONNELL, G A EAGLE, A G S MOLDAN, W D OLIFF, M J ORREN and R J WATLING (1983). South African marine pollution survey report 1976-1979. **South African National Scientific Programmes Report No. 73.** 105 pp.

KNUTSEN, J and B SORTLAND (1982). Polycyclic aromatic hydrocarbons in some algae and invertebrates from moderately polluted parts of the coast of Norway. **Water Research** 16, 421-428.

LANG, J P (1982). Environmental Pollution: Effluent discharge into the sea and estuaries. Memorandum for the Natal Town and Regional Planning Commission. 14 p and 1 map. Unpublished.

LIVINGSTONE, D J (1976). An appraisal of sewage pollution along a section of the Natal coast after the introduction of submarine outfalls. **Journal of Hygiene** 77, 263-266.

LIVINGSTONE, D J (Ed.) (1983). Detailed Report. Sludge disposal to sea. City Engineer's Department, City of Durban. September 1983. 182 pp plus a 209 pp Appendix.

LYNSKY, R (1982). **They built a city.** Durban City Engineer's Department. 1882-1982. 96 p. Concept Communications.

MACKENZIE, C R and D J LIVINGSTONE (1983). Microbial classification of seawaters. **South African Medical Journal** 64, 398-400.

McCLURG T P, B D GARDNER and N S PAYNTER (1985). Benthic macrofauna. In: **Environmental studies at Richards Bay prior to the discharge of submarine outfalls: 1974-1984.** (Eds: A D CONNELL, T P McCLURG and D J LIVINGSTONE), **Section 5.1, 45-95.**

SEGAR D A, P G DAVIS and ELAINE STAMMAN (1984). A global comparison of contamination in populated estuaries and coastal waters. Oceans '84 Conference Record. **Marine Technology Society.** p. 284-289.

SHANNON L V, P CHAPMAN, G A EAGLE and T P McCLURG (1983). A comparative study of tarball distribution and movement in two boundary current regimes: Implications for oiling of beaches. **Oil and Petrochemical Pollution** 1, 243-259.

STEBBING, A R D (1982). Hormesis; the stimulation of growth by low levels of inhibitors. **The Science of the Total Environment** 22, 213-234.

SWART, D H (1983). Physical aspects of sandy beaches - a review. In: **Sandy Beaches as Ecosystems.** (Eds. A MCLACHLAN and T ERASMUS), W Junk, The Hague. p. 5-44.

TURNER, W D (1963). In: Final Contract Report to the South African

Industrial Cellulose Corporation (Pty) Ltd. Oceanographic survey for proposed undersea outfall. Oceanographic Research Group, Durban, July 1964.

VAN EEDEN, P (1982). A survey of marine discharges along the South African coastline. **Department of Environmental Affairs Internal Report.** 65 p.

APPENDIX: NOTES ON POINT SOURCES OF EFFLUENTS

Further details are given here on the point sources of effluents dispersed to sea off the Natal coast. Note that general standards are specified for some effluents, namely:

Waste water or effluent treated to general standard shall not contain any substance in a concentration capable of producing any colour, odour or taste. pH shall be between 5.5 and 9.5. Oxygen shall be at least 75% saturation, and the effluent shall contain no typical E. coli per 100 ml. Temperature shall not exceed 35°, COD 75 mg/l, OA 10 mg/l, and conductivity 75 milli-Siemens above that of the intake water. Suspended solids shall not exceed 25 mg/l, sodium must not be increased by more than 90 mg/l above the level in the intake water. Soap oil and grease must not exceed 2.5 mg/l. In addition, the following shall not exceed the mg/l limit in the bracket: Cl (0.1); NH_3N (10.0); As (0.50); B (1.0); Cr_6 (0.05); Cr (0.5); Cu (1.0); phenol (0.1); Pb (0.1); CN (0.5); S (1.0); F (1.0); Zn (5.0); Mn (0.4); Cd (0.05); Hg (0.02); Se (0.05). The sum of Cd, Cr, Cu, Hg and Pb shall not exceed 1.0 mg/l. The waste water or effluent shall contain no other constituents in concentrations which are poisonous or injurious to humans, animals, fish other than trout or other forms of aquatic life, or which are deleterious to agricultural use (Anonymous 1984).

Table A gives details of point sources of pollution to the sea off Natal; it should be read with reference to Figure 10.3. Notes on specific outfall sites follow.

MR5: Uvongo. Incorporates a number of small discharges which previously were released to the surf zone. Now processed at the Uvongo sewage works, to general standards. Presently discharged into

Table A Details of point sources of pollution to the sea off Natal. DS = Domestic Sewage Ind = Industrial R = River

Map Reference Number	Type of Effluent	Quantity m^3/day	Plant Capacity m^3/day	Discharge Point	Comments
1. Port Edward	DS	250	-	Surf	
2. Ramsgate	DS	10	-	Surf	
3. Ramsgate	DS	160-180	228	Surf	Problems experienced during holiday season
4. Margate	DS	1 800-2 200	2 250	Surf	
5. Uvongo	DS	1 200	4 400	Vungu River	See notes
6. Shelly Beach	DS	20	-	Surf	
7. Marburg	DS	1 000	2 000	Mbango River	Discharged to river some 2 km from mouth
8. Port Shepstone	DS+15% Ind	2 000	2 046	Surf	Any excess from this work is diverted to 7 above
9. Umzimkulu River	Ind+DS	4 480	-	River	Mostly sugar mill dunder water + 10% sewage. Since 1980, treated and irrigated
10. Ifafa	DS	10	450	Ifafa River	Marina development sewage
11. Sezela sugar mill		-	-	Surf	See notes
12. Smithchem	Ind	2 750	-	Surf	See notes
13. Umzinto	DS	600	1 200	Mpambanyoni River	See notes
14. Scottburgh	DS	1 500	2 000	Surf	
15. Craigieburn	DS	150	650	Mahlongwana River	
16. SAICCOR	Ind	80 000	80 000	Submarine pipe-line	See notes
17. Umkomaas	DS	250	550	Surf	
18. Kingsburgh	DS	1 200	2 250	Little Amanzimtoti River	Has caused eutrophication problems in upper reaches of the lagoon
19. S A·Tioxide	Ind	1 700-1 800	3 300	Pipeline	See notes
20. AECI	Ind	3 000	14 000	Surf	See notes
21. Amanzimtoti	DS	12 000	20 000	Umbogintwini River	Capacity sometimes exceeded during peak holiday periods
22. Umlazi	DS	25 000	27 000	Isipingo River	Main sewage Works
23. Umlazi	DS	6000	-	Umlaas River	J Ponds
24. Durban	DS+Ind	120 000-140 000	230 000	Submarine pipe-line	Southern Works. See notes

Map Reference Number	Type of Effluent	Quantity m^3/day	Plant Capacity m^3/day	Discharge Point	Comments
25.Durban	DS+Ind	60 000–70 000	135 000	Submarine pipe-line	Central Works. See notes
26.Durban	DS	25 000	70 000	Umgeni River	Durban Heights sewage works. Purified to general standards
27.KwaMashu	DS	13 900 13 100	15 000 50 000	Umgeni River	Two plants both treating to general standards
28.Umhlanga	DS	5 500	7 200	Umhlanga River	Activated sludge and maturation pond treatment to general standards
29.Phoenix	DS	–	–	Umhlanga River	See notes
30.Umdloti Beach	DS	300–500	1 200	Surf	
31.Tongaat	DS	2 980	3 000	Hlawe River	Southern Works
32.Tongaat	DS	1 700	3 000	Tongaat River	Hambonati Works
33.Tongaat	DS+Ind	1 400	3 000	Tongaat River	Central Works
34.Tongaat	DS	50	1 650	Stream	Genezzano Works. Due to receive effluent from Westbrook in 1986
35.Whitehead	Ind	3 000	5 000	Surf	See notes
36.Ballitoville	DS	790		Surf	Now diverted to Frazers Works which has a capacity of 2250 m^3 per day
37.Sheffield Beach	DS	100		Surf	
38.SAPPI Stanger	Ind paper mill	13 000–15 000	20 000	Umvoti River	See notes
39.Stanger	DS	2 000–2 500	4 500	Umvoti River	
40.Zinkwasi	DS	10		Surf	
41.SAPPI Mandini	Ind	45 000	–	Tugela River	See notes
42.SAPPI Mandini	DS	1 300	–	Tugela River	
43.Isithebe	DS+Ind	5 000	9 500	Tugela River	Serves towns of Isithebe and Sundumbili
44.Esikhaweni	DS	3 500–5 000	5 250	Klip River	
45.Felixton	Ind	5 000	–	Surf	See notes. Board mill effluent
46.Alusaf Richards Bay	Ind	35 000	–	Harbour	See notes
47.Richards Bay	DS+Ind	130 000	240 000	Submarine pipe-line	See notes

the Vungu river, it is planned that when discharge reaches 2 400 m^3 /day the effluent will be disposed of in the surf zone.

MR11: Previously a problem area, all effluent is now recycled in the mill, including treated dunder water and domestic sewage. Stormwater drainage still reaches the beach from the offloading bays.

MR12: This effluent is mild acetic acid with traces of furfuraldehyde. Discharge is 100 m from the beach, in the inshore trough, at about 2.5 m depth (low spring tide). Initial dilutions using furfural as indicator, appear to be about 100 times within 10 m of the pipe end, on the surface under calm conditions when collecting of samples is possible (T Dreyer 1984, pers. comm.). A diving survey also revealed no apparent effect beyond a radius of about 10 m of the pipe end (T P McClurg 1984, pers. comm.). The permitted amount discharged is presently subject to an annual environmental impact study.

MR13: The MYM textile factory was discharging effluent through this works but in early 1985 this was stopped as the sewage works could not cope. The textile mill is presently commissioning a treatment works to conserve water, and about 4 m^3/day of dye, caustic soda and sodium chloride (4% solids) is being trucked for discharge via the SAICCOR pipeline at Umkomaas (MR 16).

MR16: The pipeline extends 2.16 km offshore into 24 m water depth. SAICCOR began processing timber for the production of rayon in 1956, consuming about 300 tons of timber per day. They presently process about 1 000 tons per day. Prior to 1965 effluent discharge was into the surf zone, but in that year a concrete submarine outfall was constructed capable of discharging 90 000 m^3/day. The pipeline has, from inception, functioned at maximum capacity (in reality about 80 000 m^3/day) and in view of the increased production of the plant, the effluent has become gradually more concentrated over the years. The effluent is predominantly calcium lignosulphonates in solution. Acomparison of some chemical variables is given below, comparing data for 1963 (Turner) and the present:

	pH	Temp °C	OA (ppm)	COD (ppm)	Susp. solids	TDS
1963	2,5-3,4	33-45	3 000	9 000	250	7 000
Jan. 1985	2-3	33-45	17 500	50 800	360	-

Two composite samples collected in January 1986 showed COD to have dropped to 20 500 mg/l. The toxicity and environmental impact of this effluent is dealt with separately in the text.

During early 1985 SAICCOR began construction of a new processing unit at the factory, which will be based on a magnesium rather than a calcium bisulphite digestion. The black liquor, which remains after the cooking process and removal of the cellulose fibre (used in the manufacture of rayon), contains about 18% dissolved solids, and this will be evaporated down to 55% solids. This concentrate will then be burned to generate power, with the SO_2 recovered in scrubbers. The magnesium will also be recovered from the resultant ash. The new plant will account for about 40% of factory production, thus bringing about a substantial drop in effluent concentration.

MR19: S.A. Tioxide. A highly acidic industrial effluent, discharged via a pipeline extending 1.5 km offshore to a depth of 30 m. This effluent is predominantly sulphuric acid used to separate titanium oxide from the illmenite ore. The acid contains ferrous sulphate which precipitates out of solution when the effluent is mixed with seawater. The pH of the effluent is below 1, since the sulphuric acid is about 1.2N (60-100 g/l H_2SO_4). The effluent is more dense than seawater (1.060 g/cc) and tends to remain below the surface. During 1983 a new pipe was laid with a diffuser system that was not tapered and hence in-pipe velocity and diffuser efficiency droped rapidly after the first diffuser; a low volume of discharge has added to the problem (G Toms, pers. comm.). The drought conditions of 1983-84 resulted in additional water conservation measures within the plant which resulted in reduced volume of increased strength effluent.

MR20: AECI. A mixed industrial effluent which, since water conservation measures were introduced in early 1984, has reduced in volume to less than 3 000 m^3/day. Capacity of the pipe is about 14 000 m^3/day. COD is generally about 300 mg/l, pH high (about 11), OA low at about 30, and TDS high at up to 9 000, although this latter is inconsequential in a surf zone discharge of this volume. The most problematical contaminant has been mercury, which led to some build up in accumulator organisms in the sea in the vicinity of the outfall during the late 1970s and early 1980s. Stricter control measures and water conservation in the factory have led to a marked drop in Hg levels in the effluent, and a drop in levels in marine organisms in the last few years (A D Connell, unpublished data).

MR24: Durban Southern Works. This effluent is discharged via a 4.2 km submarine pipeline of 1.37 m diameter, to a depth of 54-64 m of seawater. The pipe has a 290 m tapered diffuser section with 34 diffusers. The effluent comprises the following:

Effluent Source	Daily Volume (m^3)	pH
Sewage works discharge	100 000	6.9 - 7.0
Paper mill	30 000	5.8 - 6.6
Oil refinery (S)	4 500	8.7 - 8.8
Oil refinery (M)	2 000	7.8 - 8.8
Sugar-cane by-products	1 500	4.1 - 4.6
Sugar refinery (tanker)	100	8.3 - 8.4
Yeast factory (tanker)	100	5.1 - 5.3
Petro-chemicals effluent	25	11.7

(Connell 1983)

The final effluent mix has the following characteristics: COD of about 2 400 mg/l, OA of 420 mg/l, pH 6.5; suspended solids of 600 mg/l and TDS of 2 000 mg/l.

MR25: Durban Central Works. This sewage works, like the Durban Southern Works, is designed to remove sludge for incineration or disposal on land. However, in October 1980 experimental disposal of sludge to sea, via the Central Works pipeline, began. In June 1982 the experiment was stopped at Central Works and switched to the Southern Works. In both cases no ill effects were detected resulting from sludge disposal (Livingstone 1983), and presently disposal of sludge to sea from both plants is continuing (3% by volume). The effluent is a mixture of domestic sewage and industrial effluent, with the following approximate characteristics: COD of 1 300 mg/l, pH of 7.0; OA of 80 mg/l, suspended solids of 320 mg/l and TDS of 1 300 mg/l. The pipeline, of 1 220 mm diameter, extends 3.2 km into the sea, and is fitted with a 450 m tapered diffuser section with 18 diffusers. The pipe discharges at a depth of 48-53 m, at a rate of 60 000 to 70 000 m^3/day.

MR29: Phoenix Sewage Works. About 4 000 m_3/day is treated in oxidation ponds with a 70-day retention time, and very little effluent is thus discharged to river, but during 1986 this is expected to rise to 12 500 m^3/day.

MR35: Whitehead Textiles Effluent. Sewage from this complex was being discharged to sea with the mill effluent (1983), but is now (1985) diverted to the Tongaat Central Works. Mill effluent is discharged to the surf zone just south of the Tongaat river mouth. Characteristics of this effluent include pH of 9-11. COD of 1 000 to 3 000 mg/l, and OA of 125-250 mg/l.

MR38: SAPPI Paper Mill Stanger: Fine paper production using bagasse from the Stanger sugar mill. Current effluent characteristics include COD of 2 500 to 5 000 mg/l, OA of 300 mg/l, pH of 11.0; suspended solids 2 600 mg/l, and TDS of 1 800 to 2 000 mg/l.

MR41: SAPPI Mandini. Paper mill efluent, manufacturing cardboard from timber. Effluent characteristics include COD of 600 mg/l, OA of 150 mg/l, pH 7.0; colour of 300 to 600 Au, suspended solids of 80 mg/l, and TDS of 1 000 mg/l. Temperature is about 36°C, and conductivity about 1 200 milli-Siemens/m.

MR45: Felixton (Mondi) Board Mill. Manufacturing cardboard from bagasse, and producing a potent black effluent which has been poorly treated by an inadequately designed effluent treatment plant. Presently (early 1985) options for construction of a pipeline discharge or linking into the Richards Bay pipeline are being complicated by plans to increase (double) the capacity of the mill. At present effluent is discharged into the surf at Durnford Point (Figure 10.2), in a most unsatisfactory manner. The effluent is highly toxic (Table 10.7). Effluent characteristics include a pH of 5 to 6, COD of 10 000 to 15 000 mg/l, TDS of 7 000 to 10 000 mg/l, and suspended solids of 700 to 1 500 mg/l. Temperature 40 to 50°C. Flow rate is about 5 000 m^3/day but can be as high as 8 500 m^3/day.

MR46: Alusaf Richards Bay. Scrubbing water to capture volatile fluorine compounds is drawn from the Bhizolo canal at a rate of about 55 000 m^3/day (Figure 11.1). Due to poor flushing of this part of the harbour there has been a steady build up of fluoride in the water, such that by late 1984 intake water was generally at about 40 ppm fluoride compared with 1.5 ppm generally present in seawater. Discharge water was approximately 60 ppm fluoride. In October 1984 Alusaf began discharging 20 000 m^3/day of their discharge water to the Mhlatuzi Water Board controlled Richards Bay pipeline, while the remaining 35 000 m^3/day is being returned to the Bhizolo canal as before. A rough calculation shows that at an export rate of 20 000 m^3/day the entire holding capacity of the Bhizolo canal and Amanzimyama canals would be exchanged in less than 20 days, indicating that the present state of affairs should lead to a significant reduction in fluoride levels in the Bhizolo canal as tides introduce clean seawater to replace the loss, and fluoride is exported in the effluent discharged via the Richards Bay pipeline. The levels of fluoride in this area are being closely monitored by the National Research Institute for Oceanology in Durban (R R Sibbald, pers. comm.).

MR47: Richards Bay Double Pipeline. These two pipelines, lying side by side on the seabed, enter the sea at a point just north of the northern breakwater, at Richards Bay. The two lines have the following design characteristics:

	"A" Line	"B" Line
Construction material	HDPE	HDPE
Diameter (mm OD)	1 000	900
Design capacity (mg.l^{-1})	160 000	86 400
Total submarine length (km)	4.912	3.787
Length of diffuser section	636 m	300 m
Number of diffusers	106	16 (20)
Arrangement of diffusers	single, vertical	pairs 60° angle
Depth of water at diffusers (m)	29	26

Both lines enter a pump station which adds seawater to ensure that each line operates at full capacity at all times, although the "A" line will be capable of operating under gravity feed during power failure. The "B" line, designed to dispose of gypsum effluent, has high velocity diffusers and cannot operate by gravity. In the even of power failure gypsum will be diverted to a slimes dam. At present (early 1985) the following effluents from Richards Bay are being discharged (all data in m^3/day).

	"A" Line	"B" Line*
Alusaf	20 000 (ca 60 ppm F)	
Sewage	10 00	20 000**
Mondi	75 000	
Triomf	7 800 (ca 8 000 ppm F)	6 000 (5 200 ton gypsum/day)
Seawater	47 200	60 400
	160 000	86 400

* Started discharging from mid-June 1985
** Taken from "A" Line to form a slurry with gypsum

The major contributors were producing effluent of the following characteristics towards the end of 1985:

	Mondi	Alusaf	Triomf "A"	Triomf "B"
pH	7.0	6.1	1.2	1.8
Temperature (°C)	44	26	35	32
COD ppm	1 530	-	-	-
Total Susp. Solids	95 ppm	281 ppm	50-90 ppm	20-21%
F ppm	-	35	6 200	1 900
Volume (m^3/day)	82 500	17 500	3 600	-
Gypsum (tonnes/day)	-	-	-	2 350 - 2 900

Chapter 11

THE RICHARDS BAY MARINE DISPOSAL PIPELINES

D A Lord
Department of Oceanography, University of Port Elizabeth

G Toms
National Research Institute for Oceanology, CSIR

A D Connell
National Institute for Water Research, CSIR

THE USE OF PIPELINES FOR DISPOSAL TO SEA

The discharge of wastewater effluents into water courses such as streams and rivers or into bodies of water such as lakes, estuaries, and seas, is practiced widely. The dilution of the effluent in the receiving water, the chemical reactions which occur during the initial mixing, and the action of naturally-occurring aquatic organisms can supplement prior treatment and render wastes non-toxic.

In South Africa where wastes are discharged at most of the major coastal centres of population, 63 pipelines discharge into the sea. Of these 22 transport domestic sewage, 31 industrial effluent (14 of these are from fish processing plants), and 10 a mixed effluent (revised from van Eden, 1982). Only 10 pipelines discharge beyond the surf; two of these are less than 500 m long. The disposal of waste fresh- and seawater is controlled by the Department of Water Affairs, and discharges are regulated according to effluent standards specified by the General Standards of the Water Act of 1956. Generous exception clauses are usually applied for marine discharges and consequently, are dealt with on more of an ad hoc basis.

More recently, South Africa has developed a set of Water Quality Criteria (Lusher, 1984) for use in the control of marine discharges. The construction of the Richards Bay effluent pipeline was largely instrumental in promoting the preparation of these criteria and for the development of a more rational design philosophy for marine discharges. This design philosophy has since been applied to a proposed outfall at

Hout Bay, just south of Cape Town (Toms, 1985). Coupled with the fact that unique techniques for pipeline manufacture and laying were used in the construction of the Richards Bay pipeline, its development represents an important, interesting and novel approach to marine discharges to open coastal waters.

THE DEVELOPMENT OF RICHARDS BAY AND THE PIPELINE OPTION

Richards Bay is situated in northern Natal approximately 160 km north-east of Durban (see Figure 1.1). The decision in 1965 to develop a deepwater harbour in this estuarine embayment signaled the change from a sleepy fishing village to a modern industrial/port complex, containing the worlds largest coal trans-shipment terminal (Figure 11.1) (Lord and Geldenhuys, 1986). Coupled with this development was the encouragement for industrial growth in the area.

By 1970, two major industries were established. The first of these was an aluminium smelter (Alusaf), which produces a fluoride-containing effluent, and the second a phosphoric acid factory (Triomf) which produces 2 million tonnes per year of 'waste' gypsum, as well as a fluoride-containing effluent.

Initially the aluminium smelter effluent was treated on site to reduce the fluoride content to an acceptable level before it was discharged into an adjacent canal. The waste gypsum from the phosphoric acid factory was disposed of into slimes dams, with the fluoride being contained within these dams by sequestration with lime.

In 1979 proposals were considered for the establishment of a major Kraft type (i.e. sulphate pulping) pulp and paper mill in the area with a proposed annual production of 350 000 tonnes of fully bleached pulp. By this time the area for further expansion of the gypsum slimes dams was becoming limited, and geotechnical considerations indicated that those in operation were nearing their capacity. Consequently, impetus was given to the assessment of regional waste disposal requirements, and specifically into the development of a joint wastewater disposal system (DEA, 1981), which meant a large marine outfall or pipeline.

The option for such a pipeline to be operated as part of a joint wastewater disposal scheme for the area was not new, as this concept

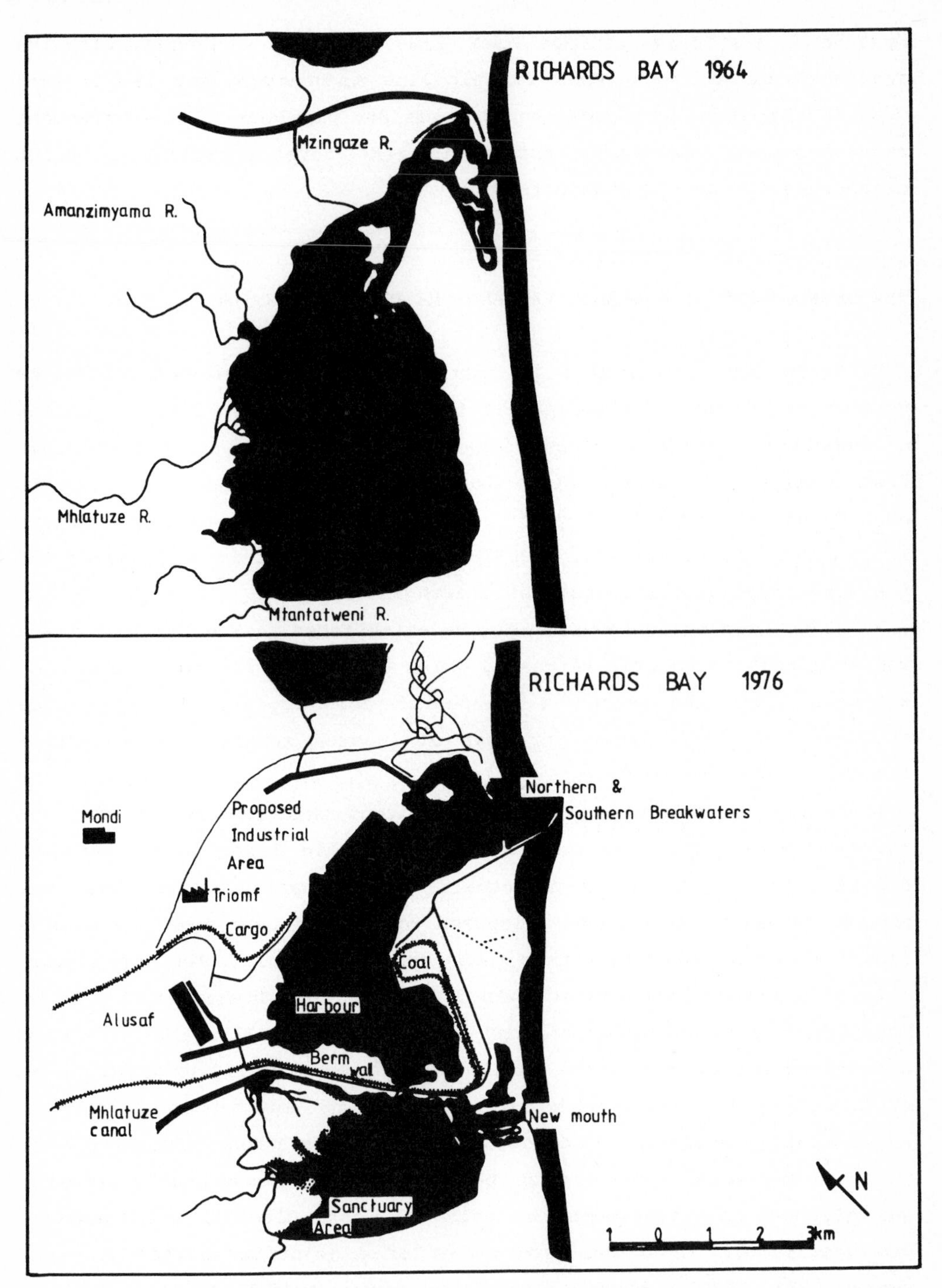

Figure 11.1 Richards Bay 1964 and 1976. The changes made in the development of the port are clearly evident. A southern sanctuary area with a new mouth is kept separate from the harbour by means of a berm wall, while the industrialised areas, coal terminal and canalized rivers are shown.

had first been proposed in 1969. Since then reasonably extensive but fairly broad oceanographic data had been collected within the general area (Pearce, 1977), while there had also been specific measurements made to assist in the design of the Richards Bay harbour (NRIO, 1975 etc). All of these data were subsequently summarized (NRIO, 1981) and used in the detailed design of the pipeline. This detailed planning and design included the following major considerations;

a) assessment of the marine environment (physical, chemical, biological);
b) effluent characterization, and identification of critical constituents;
c) the determination of discharge criteria;
d) the design and construction of the pipeline system; and
e) monitoring after the start of operation.

ASSESSMENT OF MARINE ENVIRONMENT

Physical

The sea bed off Richards Bay slopes relatively gradually to the east and more gradually to the south (Figure 11.2). To the east, the 20 m contour is 4 km offshore and the 30 m contour is 5.5 km offshore; to the south, these isobaths are 7 and 12 km offshore respectively. The inshore boundary of the Agulhas current is usually situated about 20 km offshore (approximately at the 300 m isobath), with this edge meandering at distances from 10 to 30 km offshore (Pearce, 1977).

Coastal currents are predominately wind driven (Schumann, 1981) and this area is one of generally low current velocities, although values of 65 cm/sec have been measured. Detailed current data collected over a two- year period (NRIO, 1981), at a point to be approximately at the end of the pipeline, (depth 17 m, information collected 2 m above sea floor), showed that calm periods, which are considered to be those when current velocities are less than 3 cm/s, occur up to 50% of the time. Consequently average near-bottom current strengths in the area are low, at 8.5 cm/s. The persistence of higher velocity currents is limited, with near-bottom current speeds of 50 cm/s occurring at most for 2.5

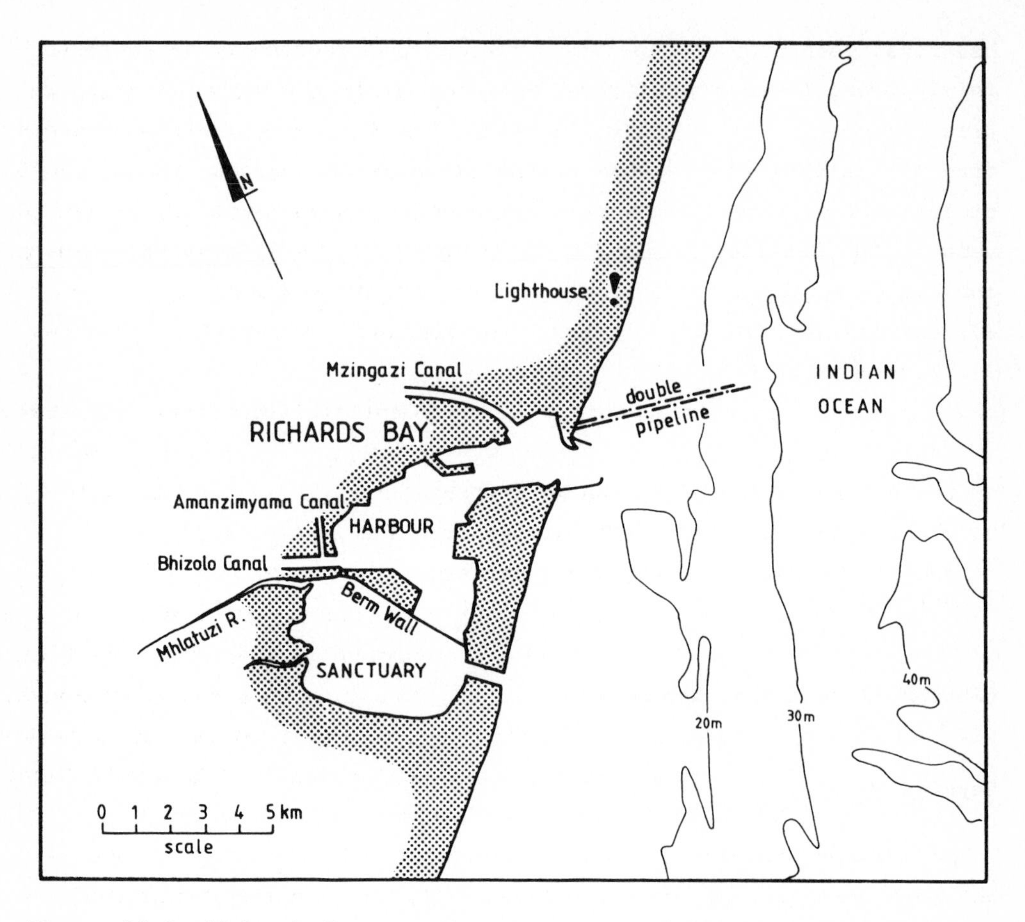

Figure 11.2 Richards Bay, on the east coast of Natal. Bathymetry is shown as well as general pipeline routing; the longer buoyant outfall lies to the south.

hours at a time. In comparison, calm periods sometimes extend for three days. The dominant current directions are ENE-ward (current flows towards the ENE) and southwestward. The currents flowing ENE-ward are stronger (in sympathy with the stronger SW winds). However, southwestward-flowing currents occur more frequently. Current reversals tend to occur every 2 to 5 days (NRIO, 1981).

Data collected since 1968 show that 90% of wave heights range from 0.5 m to 2.0 m, with the majority of wave periods ranging from 8 to 13 seconds. The majority of wave directions fall within the sector south to east, with the predominant direction SE (NRIO, 1981).

Biological & Chemical

Prior to construction of the Richards Bay pipeline, relatively detailed biological and chemical studies were made in the vicinity of the proposed outfall during the period 1981 to 1984 (Connell et al., 1985). The interpretation of monitoring data was complicated by the fact that the construction and operation of Richards Bay harbour had required extensive dredging. Prior to 1983, several million cubic metres of dredge spoils were dumped in the general area destined to receive the pipeline discharge (McClurg et al. 1985). Apart from the effects of the disposal of these dredge spoils which are uncontaminated, the region is largely undisturbed and 'unpolluted'.

In addition, floating rafts have been moored in the vicinity of the proposed pipeline as well as at comparative or 'control' sites further away. These rafts are equipped with removable PVC settlement plates, and allow for a time-series analysis of the fouling communities at these sites (van der Elst, 1983).

EFFLUENT CHARACTERISATION

Major components of concern

Detailed pipeline design was based on effluent characteristics provided by the four main bodies which intended to use the pipeline, namely, Alusaf, who operate an aluminium smelter; Triomf, who operate a phosphoric acid plant; the Richards Bay Town Board who were proposing to use the pipeline for the co-disposal of town sewage; and Mondi, who proposed to erect a pulp and paper mill (Table 11.1). A study of each of the proposed effluents from these then allowed the potentially harmful components of each of the effluents to be identified (Table 11.2). It was concluded that the undissolved gypsum and the dissolved fluoride presented the greatest problem. This was based on chemical measurements as well as on acute and chronic bioassays of actual and simulated effluents (Connell et al. 1985). To finalise the design of the pipeline and its performance, maximum allowable levels of undissolved gypsum and dissolved fluoride were then established.

Table 11.1 Effluents proposed for incorporation into the Richards Bay pipeline (DEA 1981). COD = Chemical Oxygen Demand, TDS = Total Dissolved Solids, SS = Suspended Solids.

Source	Triomf Fertilizers	Mondi Paper Co.	Richards Bay Municipality	Alusaf	Total
Volume (m^3/day)					
Initial	25 800	90 000	5 000	20 000	140 800
Future	25 800	160 000	45 000	20 000	250 800
Main Constituents	Gypsum 6 440 tonnes/d fluorides 92 tonnes/d heavy metals	Organics (COD,TDS, SS) colour	Organics (COD, SS)	Fluorides	

Table 11.2 Major environmental concerns with proposed effluents.

Item	Concern
Gypsum	Large volume, cloud effect (aesthetics and clogging of fish gills), smothering effect on ocean floor.
Fluorides and heavy metals	Acute toxicity and sub-lethal effects on biota.
Biodegradable organic materials	High oxygen demand.
Chlorinated organics	Effect on biota, accumulation in the food chain.
Colour	Aesthetics and reduced light penetration.

Gypsum Disposal

Gypsum is produced as a byproduct in the manufacture of phosphoric acid, by reaction of sulphuric acid with phosphate rock (apatite). This gypsum (calcium sulphate dihydrate i.e. $CaSO_4$ $2H_2O$) is frequently referred to as "phosphogypsum", and has been produced at Richards Bay at the rate of 2 million tonnes per year. Its disposal at Richards Bay thus far has been in slimes dams.

Phosphogypsum is moderately soluble in seawater (0.15% by weight). Once dissolved it is totally harmless as calcium and sulphate are two of the major ions in seawater. The rate of dissolution of gypsum is relatively slow, and like most salts, this increases with temperature. To obtain optimum dissolution of the gypsum in seawater, the pipeline design required that first, the gypsum was diluted to a concentration below 0.15% by weight in the water column and second, that the gypsum particles were suspended in the water column for long enough at this or less than this concentration to allow them to dissolve while settling. Reliance was then placed on ambient mixing (wave motion, currents, diffusion) to enhance further dissolution of any particles reaching the sea floor (NRIO, 1982).

Fluoride

Dissolved fluoride in the effluent originates mainly from the manufacture of phosphoric acid from the fertilizer factory. Fluoride occurs as an impurity in the phosphate rock used as raw material. The bulk of this fluoride is dissolved during the reaction with sulphuric acid, and is then removed during the process of concentration of the phosphoric acid.

Fluoride is known to be toxic to fresh water organisms at levels as low as 1 to 2 mg/ℓ. Some international literature suggests the same levels constitute a hazard to the marine environment. However, coastal seawater naturally contains 1.3 to 1.7 mg/ℓ of fluoride.

This lack of international norms for acceptable levels of dissolved fluoride in sea water necessitated an experimental bioassay programme to determine acceptable fluoride levels, which was implemented using commonly occurring local organisms. Of the many tested, the widely occurring burrowing amphipods (Grandidierella lutosa Barnard and G. lignorum Barnard) proved to be of the more sensitive to fluoride, but also very convenient for conducting multigeneration chronic bioassay tests (Connell and Airey, 1982). Using reproductive success of these organisms as the indicator of acceptable conditions, it was assessed that sub-lethal toxic effects could be expected at fluoride levels exceeding 5 mg/ℓ. For the Richards Bay discharge, these data have been translated into the permit requirement to mean that the level of 5 mg/ℓ must not be exceeded outside the mixing zone of the effluent plume. In this manner the discharger is allowed to use the initial dilution gained upon discharge of the effluent from the marine outfall. The mixing zone is then limited to the zone where initial rapid dilution of the effluent is occurring i.e limited by the water surface above the diffuser and, by definition in Lusher (1984), as a volume equal in depth to the depth of water over the diffuser, and in width to twice the depth of water plus the width of the diffuser, and in length to twice the depth of the water plus the length of the diffuser.

DESIGN AND CONSTRUCTION OF RICHARDS BAY PIPELINES

Design concepts

In its initial form, the principle of the design of the Richards Bay pipeline consisted of combining all the effluents and discharging these through a single pipeline. This design concept included a very important compromise limiting the optimum performance of the pipeline. The compromise results from the fact that to discharge the buoyant part of the effluent containing the fluoride in the most effective way, it is advantageous to discharge as deeply as possible to take advantage of the greater depth of water to obtain greater dilution. To achieve maximum dilution by buoyant rise is best done by the use of a deep diffuser provided with a large number of ports that distribute the effluent into as many rising buoyant plumes as possible. Consequently the exit velocity from the ports in this case are usually low. However, the major requirement in discharging dense or negatively-buoyant effluents is to discharge them at a high exit velocity to encourage jet entrainment. This is most easily achieved by forcing the discharge through only a few ports. Where a single effluent contains separate components with widely different characteristics, the effluent can be discharged at an intermediate depth where some wave action is still present on the sea bed to enhance mixing of the dense effluent, but where the depth is still sufficient to provide adequate dilution.

After a thorough review of the components of the proposed pipeline, and in recognition of the compromise required, it was decided to separate the wastes at, or close to, their source into a dense mixture and a buoyant mixture. Each of these could then be discharged through separate outfalls and diffuser systems (Toms 1986; Roberts and Toms 1986).

Discharge of buoyant effluents

Marine discharges (or ocean outfalls) are of value mainly because of the large dilutions which can be rapidly achieved with buoyant effluents. Effluents are regarded as being buoyant if their densities are lower than that of the surrounding seawater. This is usual as fresh water is the major component of most effluents.

Upon discharge of an effluent from a submerged outfall there is rapid initial dilution followed later by advection, diffusion, and decay of components in the effluent which reduces concentrations further (Russell 1984).

Initial dilution results in the rapid turbulent mixing of waste water with ocean water around the point of discharge. For a submerged buoyant discharge, the momentum of the discharge and its initial buoyancy act together to produce turbulent mixing. Initial dilution in this case is completed when the diluting waste water ceases to rise in the water column and begins to spread horizontally.

The amount of initial dilution which occurs is strongly influenced by the density difference between the effluent and the surrounding seawater, as well as by the vertical density gradient of the surrounding seawater column through which the buoyant effluent will rise. In addition, initial dilution is influenced by water depth, movement of the surrounding seawater (i.e. the presence of currents), the exit velocity of the effluent from the outfall pipe, and the size and spacing of the diffuser ports along the outfall pipe. Procedures for the prediction of such initial dilution - which is under the control of the designer - are well developed (Abraham, 1963; Fan and Brooks, 1969; Roberts, 1977).

Discharge of negatively-buoyant effluents

The discharge of negatively-buoyant effluents by pipeline is a less common practice, with effects and design criteria normally assessed on a case by case basis. Obviously such effluents will not enjoy any dilution due to buoyant rising in the water column. Instead, substantial initial dilution can only achieved by discharging the effluent into the water column at a high velocity. This will allow for jet entrainment at the shear interface between the fast moving effluent jet and the relatively still ambient seawater, followed by the decay of components of the effluent by decomposition and especially dissolution.

Design Details

The buoyant effluent line was designed to transport the bulk of the Mondi pulp-mill effluent together with that portion of the wastewater from the Triomf fertilizer plant containing up to 92 tonnes/day of fluoride, devoid of gypsum. The design criterion for this pipeline was not to exceed a concentration of 5 mg/ℓ of fluoride in the seawater beyond the mixing zone. A two-hundred fold dilution of the effluent would allow this criterion to be met. Using standard procedures for calculating dilution (Wright 1984), it was shown that this dilution level could be achieved easily with a pipeline having a diffuser section with 106 ports at 6 cm centres in a water depth of 29 m (5 km offshore). Under stagnant conditions (zero current, no stratification), such a diffuser section was estimated to achieve an average dilution of 270 times.

Currents would increase dilution markedly, while stratification of the water column would prevent the plume surfacing and thereby decrease dilution, but would also make the plume invisible. For example, density stratification due to a 2°C difference in temperature in the water column would limit the plume's rise to within 5 m of the surface and would reduce dilution to only 200 times. A 10 cm/s current imposed on this would further reduce the plume rise to 10 m below the surface, but would increase the dilution to 820 times.

The dense effluent line was designed to transport all of the gypsum as a slurry using a small amount of pulp-mill effluent, to a maximum of 6400 tonnes per day. To obtain maximum dissolution of the gypsum in seawater, the pipeline design required that first, the gypsum is diluted to a concentration below 0.15% by weight in the water column and second, that the gypsum particles are suspended in the water column for long enough at this or less than this concentration to allow them to dissolve while settling. Reliance is then placed on ambient mixing (wave motion, currents, diffusion) to enhance further dissolution of any particles reaching the sea floor (NRIO, 1982).

This is shown schematically Figure 11.3. Theoretical worst case dilution for this line is 170 times. This is the average dilution across the dense effluent jet when it reaches the sea floor at the end of its trajectory. At the apex of the plume, the dilution at the centre-line is 50 times, while the average dilution across the effluent jet at the apex is 85 times. The design recommended that the slurry containing undissolved gypsum be expelled from 16 ports at the very high velocity of 15 m/s. The ports were arranged in eight sets of two each, the two ports of each set being on opposite sides of the pipe and inclined at 60° to the horizontal. The end of this pipeline is located 4 km offshore at a depth of 25 m and the trajectory of the top surface of the plume of the discharged material is predicted to reach a height of 15 m above the sea floor. This will allow each particle to be suspended in the water for longer than 25 minutes (when slow dissolution can occur), which coupled with the dilution expected, will allow for more than 95% of the gypsum to dissolve before reaching the sea bed.

The Use of HDPE (High Density Polyethylene) Pipes

Traditionally marine outfalls are constructed from iron, steel, concrete coated steel, or reinforced concrete (Grace, 1978; WHO, 1982; Moss-Morris, 1984). More recently various polymers have also been used for pipelines, particularly high density polyethylene (HDPE). These 'plastic' pipes have the advantage of being lighter and more flexible than conventional materials, which makes their laying far simpler. In addition, they are resistant to corrosion, and far cheaper to construct. Their reliability is unknown as such outfalls have not been in service for sufficiently long periods of time. For the Richards Bay lines, the selected material of construction was HDPE. The use of HDPE for such long outfalls into relatively hostile open coastal waters (where the design 1 in 10 year wave height is 12 m) is rather unique and represents one of the more novel aspects of this marine disposal

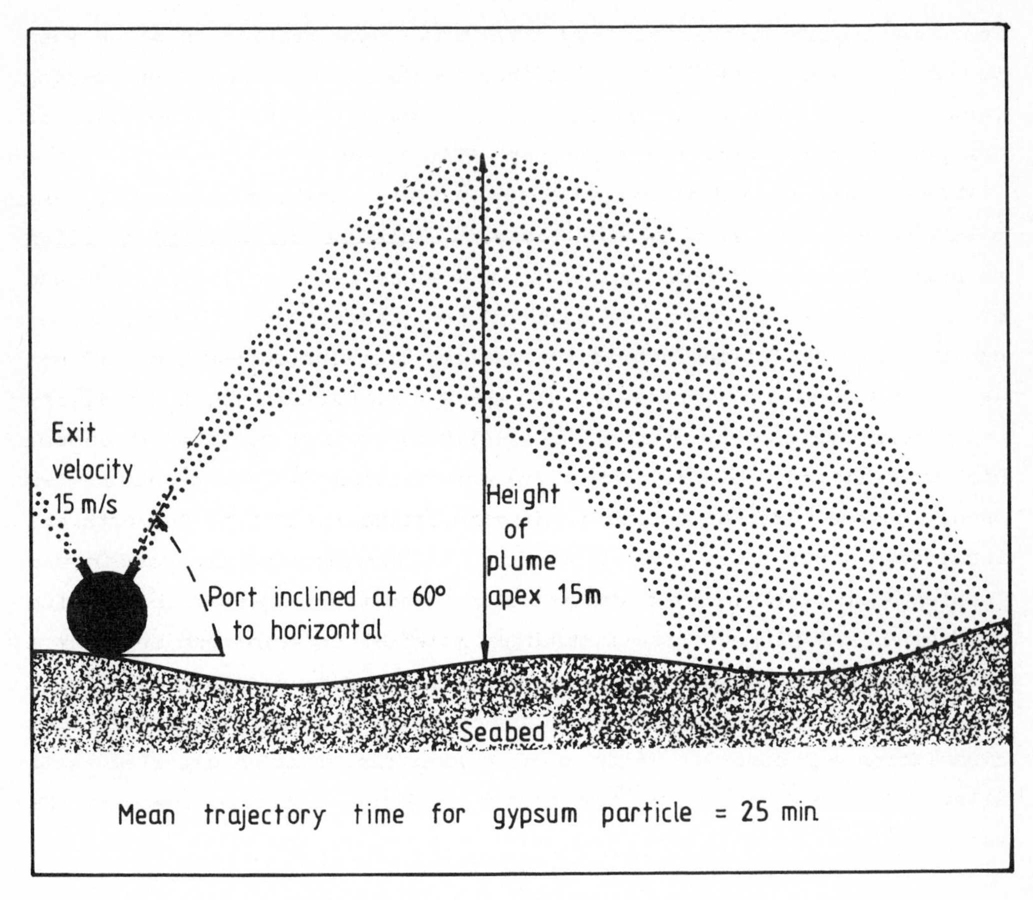

Figure 11.3 Discharge of gypsum.

project.

In its final form the scheme consists of two pipes, one to discharge a buoyant effluent, the second to discharge the dense effluent. The sea-lines consist of pipes of 1 000 mm (5 km long) and 900 m (4 km long) external diameter respectively, each having a 40 mm wall thickness. The pipe was to be placed on the seabed (not buried) and weighted at regular intervals with concrete collars. An interesting aspect of the design was that the determination of the required weighting was based on the assumption that the pipe would be supported by star-shaped anchor weights which would keep the pipe free of the sea-bottom, thereby reducing horizontal and uplift forces. Furthermore, because the HDPE pipe is flexible the design could be based on a design wave height of 1 in 1 year, rather than say a 1 in 100 year.

This means that the HDPE pipe is allowed certain movement over the sandy sea-botom under extreme sea conditions. The pipe routes also crossed two short reef areas and weighting needed to be increased in these to reduce movement which could lead to abrasion and possible failure. It was also recommended that weighting be increased close to the diffusers. The specific design technique for flexible pipelines (Abbott et al., 1977) required further development locally as reported by Pos et al., (1986).

Pipes were manufactured on site by direct extrusion in lengths varying from 400 m to 600 m. In passage through the surf zone the pipes were weighted and buried beneath the sea bed. The preparation of the sea bed in this area for the pipes required the construction of a 400 m long jetty which was dismantled after the laying was completed in June 1984.

Pipeline operation

The buoyant line first started discharging in October 1984, and by June 1986 the paper mill was contribution about 85 000 m^3/day to it. Due to a series of failures in the landline section of the high pressure, dense effluent line, discharge only began in July 1985. By May 1986 Triomf was operating at about 80 per cent capacity and by the end of May was operating at full capacity although the discharge from this line has been no more than half of the design rate. During the survey of May 1986 no gypsum was encountered in the grab samples from an array of stations around the pipe ends, but divers reported that gypsum was collecting around the last two pairs of ports of the diffuser and had also accumulated alongside the buoyant line for about 100 m to seaward of the end of the dense effluent line diffuser (A D Connell, unpublished). By mid-1986, gypsum discharge was terminated due to the Triomf factory stopping production. It is expected that production, and therefore discharge, will recommence in the near future.

MONITORING OF THE RICHARDS BAY PIPELINE

Monitoring and assessment of the pipelines after operation commences will concentrate on determining how closely the dilutions achieved agree with predicted levels as well as on ascertaining the degree of any environmental disturbance.

Features influencing the behaviour of the buoyant effluent plume, such as waves, currents, and vertical stratification will be routinely measured, while the position of the effluent plume in the water column will be assessed using natural or artifical tracers.

Effluent from the dense pipeline will be affected by ambient conditions to a lesser extent than that from the buoyant line due to the much higher effluent exit velocities from this line. Performance monitoring will include elucidating the shape of the plume trajectory, and the build-up of undissolved gypsum.

Chemical monitoring will concentrate on those materials in the discharge which are regarded as potential problems. Monitoring of fluoride has already commenced, and recent data (June 1986) show that no marked increase even in surface samples taken from within the visible plume of the buoyant effluent. In addition, copper concentrations in both water and sediment will be followed as significant amounts of copper are present as impurities in the rock phosphate used in the manufacture of phosphoric acid. The paper mill effluent will be followed using measurements of dissolved organic carbon.

Biological monitoring at sea includes studies of benthic organisms from a grid of stations around the end of the pipeline, beam trawling and neuston netting, the latter used for fish egg and larvae studies. Beam trawled specimens of bottom-dwelling fish such as soles and flounders, and some crustaceans, are used for trace metal analysis, and the livers of flat fishes are sectioned for histopathological studies. Bacteriological sampling from the vicinity of the pipeline has also been continued, and as yet no elevated counts have been found (June 1986; D J Livingstone, pers. comm.).

Along the beaches adjacent to the pipeline, meiofauna studies, some limited chemistry of interstitial waters, and surf water bacteriology studies continue, as described in Connell _et al_. (1985). Meiofaunal communities have been severely reduced by dredge-spoil dumping on the beaches (Whitehorn, 1985) and the pattern of recovery has been followed. Abundance and diversity of meiofauna reached satisfactory levels some 18 months after cessation of dumping.

Acknowledgements

The authors would like to thank the SANCOR programme of the CSIR, and the Department of Environment Affairs for support for some of the work reported. In addition, we thank the Mhlatuze Water Board and their engineering consultants for permission to publish information concerning details of the pipeline scheme.

REFERENCES

CONNELL A D, T P McCLURG and D J LIVINGSTONE (1985). Editors: Environmental studies at Richards Bay prior to the discharge of submarine outfalls: 1974-1984. **Marine Research Group,** NIWR, CSIR, 289 pp.

CONNELL, A D and D D AIREY (1982). The chronic effects of fluoride on estuarine amphipods *Grandidierella lutosa* and *G. lignorum*. **Water Research** 16, 1313-1317.

DEA, (1981). **Report on the proposed Richards Bay sea outfall scheme.** Department of Water Affairs, Forestry and Environmental Conservation. Government Printer, Pretoria, 26pp.

EPA, (1982). **A Compilaton of Water Quality Standards for Marine Waters. EPA,** Washington D.C.

GRACE, R A (1978). **Marine outfall systems.** Prentice-Hall.

LORD, D A (1984). The case for water quality in South African marine waters. In: **Pipeline discharge of effluents to sea** (Eds: D A Lord, et al.) SANSP Report No. 90, CSIR, Pretoria.

LORD, D A and N C GELDENHUYS (1986). The Richards Bay effluent pipeline. **SANSP Report No. 129,** CSIR, Pretoria.

LUSHER, J (1984). Water quality criteria for the South African coastal zone. **SANSP Report No. 94,** CSIR, Pretoria.

McCLURG T P, B D GARDNER and N S PAYNTER (1985). Benthic macrofauna. Report 51 In: **Environmental studies at Richards Bay prior to the discharge of submarine outfalls: 1974-1985.** (Eds: Connell, A.D. *et al.)* Marine Research Group, NIWR, CSIR. 289 pp.

MOSS-MORRIS A, (1984). Submarine Pipelines: Practical considerations in their design and installation. In: **Pipeline discharge of effluents to sea** (Eds: D A Lord *et al.*) SANSP Report No. 90. CSIR, Pretoria.

NAS, (1973). Water Quality Criteria 1972. National Academy of Sciences, National Academy of Engineering, U.S. E.P.A. Washington, D.C.

NEL, P R C (1981). Input-output of by-product gypsum. **Report INFO 110.** I R S, CSIR, Pretoria, 20 pp.

NRIO, (1975 etc). Richards Bay Harbour Field Studies - Series of Reports submitted to South African Transport Services by National Research Institute for Oceanology, Stellenbosch.

NRIO, (1981). Richards Bay Ocean Outfalls. Final Report. Report No. C/SEA 8116, National Research Institute for Oceanology, Stellenbosch.

NRIO, (1982). Richards Bay Ocean Outfalls. Engineering Design Aspects. Vol. III. Dilution Calculations and Design of the Diffuser. November, 1982.

PEARCE, A F (1977). The shelf circulation off the east coast of South Africa. **NRIO Professional Research Series Report No. 1.** 220 pp.

POS J D, K S RUSSELL and J A ZWAMBORN (1986). Wave force and movement calculations for a flexible ocean outfall pipeline. Paper accepted for presentation at 20th ASCE International Conference on Coastal Engineering, Taiwan, November 1986.

ROBERTS, P J W and G TOMS (1986). Inclined dense jets in a flowing ambient. Paper submitted to ASCE **Journal Hyd. Eng.** for publication.

RUSSELL, (1984). Engineering design in pipeline discharges of effluents to sea. In. **Pipeline discharge of effluents to sea** (Eds: D A Lord et al.) SANSP Report No. 90, CSIR, Pretoria.

SCHUMANN, E S (1981). Low frequency fluctuations off the Natal coast. **Journal of Geophysical Research,** 86, 6499-6508.

TOMS, G. (1985). Application of the report 'Water Quality Criteria for the South African Coastal Zone' to the Hout Bay Outfall Project. NRIO Report T/SEA 8504/2. Stellenbosch, 24pp.

VAN DER ELST R P, W O BLANKLEY and S CHATER, (1983). Richards Bay Rafts. Presented at 5th National Oceanographic Symposium, Grahamstown, January 1983.

VAN EDEN, P G (1982). A survey of marine discharges along the South African coastline. Report issued by Department of Water Affairs, Water Pollution Control Division, Pretoria. April 1982.

WHITEHORN, J D (1985). Beach meiofauna. In: **Environmental studies at Richards Bay prior to the discharge of submarine outfalls 1974-1984.** (Eds: A D CONNELL, T P McCLURG and D J LIVINGSTONE), Report S2: 97-145.

WRIGHT, S J (1984). Buoyant jets in density stratified crossflow.

ASCE, **Journal of Hydrological Engineering,** 110(5), May 1984.

WHO, (1982). **Waste discharges into the marine environment. Principles and guidelines for the Mediterranean Action Plan,** WHO, UNEP, 422 pp.

Appendix

THE R V MEIRING NAUDÉ

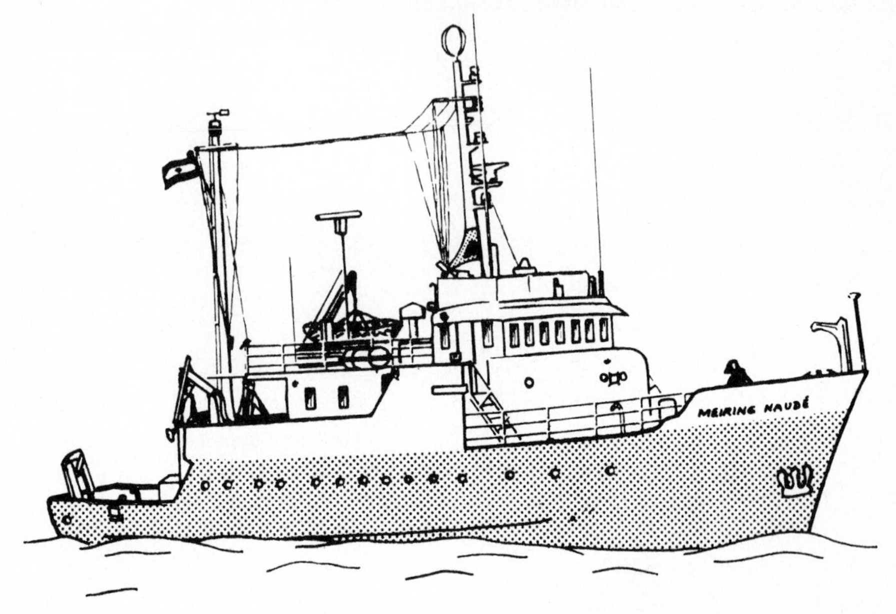

A line drawing of the RV Meiring Naudé

The Research Vessel Meiring Naudé was used for many of the investigations reported in this volume. A steel hulled vessel designed specificaly for oceanographic research, she is owned and operated by the National Research Institute for Oceanology of the Council for Scientific and Industrial Research.

The RV Meiring Naudé was built in 1967 by the Barens Shipbuilding and Engineering Corporation Limited, in Durban. Specifications are given below.

Length	Beam	Draft	Displacement
31.75 m	7.62 m	2.97m	364 tonnes

The vessel has a maximum speed of 10 knots, but can maintain a cruising speed of 9 knots. She has a fuel range of 4 000 nautical miles, an endurance of 12 days, has a crew complement of 13 and can carry up to 8 scientific staff.

An MK21 Decca Navigator, as well as a Navidyne ESZ 4000 satellite navigator are available for navigation. An electro-hydraulic winch, with a capacity of 5 000 m of 6 mm electro-mechanical wire is used for the underwater instruments, while a portable winch with a capacity of 5 000 m of 9 mm wire is also available.

An anti-roll tank, as well as a 75 HP Pleuger electric bow-thruster are fitted. The vessel has been extensively treated (anti-vibration mounts/acoustic panelling) to reduce noise.

Laboratory space consists of the following:

a) Wet laboratory (16.3 m^2) has hot/cold, fresh/sea water supplies, storage space for bottles, and a 600 ℓ deepfreeze. A continuous supply of sea water is available from an intake near the bow.

b) Electronics laboratory (15 m^2) has a static inverter supplying 2KVA 220V, 50 Hz precision frequency and stabilized voltage.

c) Biological/Chemical labratory (15 m^2).

The dry laboratories (b and c) have separate non-recirculating air conditioning. There is an open deck aft, of 36 m^2, as well as deck space of 5.5 m^2 adjacent to the wet laboratory.

REFERENCE

STAVROPOULOS, C C (1985). The Research Vessel "Meiring Naudé". NRIO Data Report D8503.